# ADVANCED DREDGING TECHNIQUES

## PART 1

## Finding and Recovering Pay-streaks

By

Dave McCracken

# BY THE SAME AUTHOR OF:

**GOLD MINING IN THE 21st CENTURY**
The Complete Book of Modern Gold Mining Procedures $19.95

**BASIC GOLD DREDGING TECHNIQUES**
How to Find a lot of Gold Using a Suction Dredge $10.95

**ADVANCED DREDGING TECHNIQUES, PART I**
Finding and Recovering Pay-streaks $9.95

**ADVANCED DREDGING TECHNIQUES, PART ll**
Succeeding at a Gold Dredging Venture $9.95**VIDEOS**

**VIDEOS**
**(Available either in VHS or DVD format)**

**MODERN GOLD MINING TECHNIQUES** $49.95

**SUCCESSFUL GOLD DREDGING MADE EASY** $49.95

**ADVANCED DREGDING & SAMPLING TECHNIQUES**
$49.95

Available from Keene Industries
20201 Bahama Street, Chatsworth, California 91311
1800 841-7833 or 800 392-4653 from California)

Autographed copies available from New Era Publications
27 Davis Road, Happy Camp, California 96039
(530) 493-2062

**www.goldgold.com**

For more information about the author and his ongoing programs, either visit **www.goldgold.com** or write to him at :

Dave McCracken
27 Davis Road/P. O. Box 153
Happy Camp, California 96039

PUBLISHED BY KEENE INDUSTRIES
20201 Bahama Street
Chatsworth, CA 91311
(818) 933, 0411

**www.keeneengineering.com**

# Table of Contents

# Acknowledgement

In early 1980, while I was just getting my own start in mining, I sent a letter to Jerry Keene telling him that from the knowledge I had gained while supporting myself with a 4-inch dredge, I believed that I could put together a book, and call it ***"Advanced Dredging Techniques."*** I explained in my letter that I believed what I had learned about hard-packed natural streambeds, lines of gold, pay-streaks and sampling procedures could be very helpful to others. At that time, the technology of how to sample for high-grade underwater gold deposits did not appear to be in any book or manual.

I was not much more than a kid in those days, living in a tent out on the river, looking for a place to make my own mark on the world. I didn't have much more than an idea, and I didn't even expect to receive an answer from Mr. Keene. But I got a letter back from him just a few weeks later saying that he liked my idea about writing an advanced manual on suction dredging.

Had I not received an encouraging response from Jerry Keene, it is likely that I never would have written that first book. The first book was followed by others and some videos. Those helped me build my own foundation in the small-scale mining industry. The years have gone by (faster than I like to think about) and I have built my whole life around small-scale gold mining.

During the past 50 years, there have only been a few individuals or companies that have had a profound affect on shaping the small-scale mining industry. Of those, no other company has had a more meaningful impact than the Keene family; three generations of quality service to keep the golden dream alive for everyone.

The Keene family legacy in gold mining was started by Jerry's father, Ernie Keene. Ernie developed some of earliest underwater suction dredges during the 1950's and used them extensively along the Yuba River in Northern California. Those dredges opened access to rich underwater gold deposits that had never been accessible before. Word got around fast about Ernie's rich discoveries. Soon thereafter, Ernie's friends and many other miners started placing orders; and dredge-building developed into a full-time business for the Keene family.

Jerry spent countless hours mining and working in the shop with his dad. Using his business and organizational skills, Jerry developed the operation into a million-dollar business, turning Keene Engineering into the world's largest supplier of portable mining equipment.

Jerry's two sons Pat and Mark were involved in the family business even as children. They have played a <u>major</u> roll in developing the modern equipment which you see in the field today – all across the world. As a result of their continuing hard work, KEENE is the industry standard which all other portable mining equipment is measured against. You will see

Mark and Pat at nearly all of the industry trade shows, continuing to promote the industry, and communicating their own excitement about gold mining to the next generation of small-scale gold miners who are just getting started.

Looking back, I wonder where I would be today if I had not received those first words of encouragement from Jerry Keene. I am nearly certain that my path would have led elsewhere. It's funny how just a few encouraging words from someone you look up to can change your whole life. We should always remember this, because it is one important way that we leave our mark on the world.

*To Maria,*

*the best and most loyal friend*
*I have had in this life.*

# Introduction

A short time ago, my dredging partner and I were sampling a new section of river. We were dredging test holes, searching for high-grade pay-streaks, and hadn't had any luck from five very well-done samples. We floated down to a new section of river where the water was very fast. It was on the inside of the tail-end of a bend of the river - an excellent place to find a high-grade pay-streak.

The problem was that we had not found a single speck of gold in the previous five sample holes. This was uncommonly-bad! Usually, we at least find some showing of flood-gold in every test hole in river-dredging. For the most part, I had decided that this section of river must have been very low-grade. It was about a half-mile downriver to the next river access where we would be able to pull the dredge back out of the river. So we decided it was worth a few more sample holes as we drifted down in that direction.

This new section of river had slow water towards the bank. I had it figured that the pay-streak would be located out under that fast water - which is often the case. I don't know why, but mother-nature often has a knack for hiding her richest treasures in areas which are most difficult to get into!

We set up the dredge in the slower water alongside the bank, got everything running, and I started the sample hole right out on the edge of the fast water.

Once the hole was down far enough to protect me from the current, I pushed the out under the fast water where I figured the pay-streak would most likely be. This was very difficult, because the force of water was pushing very hard against the suction hose and my airline. I just muscled the work through against the current, but it took a great deal of effort. I was in that zone where heavy effort, physical pain, and emotional stress are all one and the same.

The suction hose was swept back into the current several times. This required me to go back, drag the hose over towards the bank, pull it back upstream, and get it back out into the sample hole. All this, to keep pushing the hole further out into the current, and deeper towards bedrock.

When I reached bedrock, I uncovered it slowly to see if there was any visible gold in the cracks. All I saw was a few small flecks - nothing to get excited about. It was gold, however; and seeing it on the bedrock was encouraging. This was the <u>first</u> gold we had seen in that section of river. So it was a good reason to investigate further.

Having it in mind that the best gold would be further out into the current, I pushed the hole in that direction. The further I pushed, the more difficult it was. The bedrock out there showed no gold. My arms felt like spaghetti!

So I pulled back into the original place where I spotted the gold flecks on bedrock, took another small cut off the front of the hole, and worked it down to bedrock slowly so I could see if there was any more gold. I did not see a single speck! So I decided this area must also be low-grade and decided to call it quits.

As I was dragging the dredge's suction hose in towards the bank, the thought crossed my mind for the first time that I ought to test the inside of the dredge hole - over towards the slow water.

If the streambed material had been deeper, I probably would not have tested toward the inside. Why? Because I had it in my mind that the pay-streak was supposed to be out under the fast water. But, because the streambed was shallow, and I had seen some gold on the bedrock where I first touched down, I decided to take a quick cut off the inside of the sample hole.

Sure enough, I spotted several small flakes of gold on the bedrock. So I took a larger cut off the inside, pushing towards the bank. About halfway down to bedrock, I started seeing flakes of gold in the gravel - kind of like Christmas. On the bedrock, I uncovered a pothole full of gold. It was about eight ounces, and about half of it was jewelry gold. And that began one of the richest pay-streaks we ever found!

There is a lesson in this: Everything we do and everywhere we go is the result of the decisions we make. Sometimes we get going on a path which is just not the right direction! This is not just in gold mining. Sometimes the signals are there, telling us we are going the wrong way - often by the amount of pain and discomfort and effort it is taking to keep moving in the wrong direction.

Have you ever found yourself in a major difficulty - only later, sometimes much later, to realize that it was one of the best things that ever happened to you?

***The lesson is to never give up hope!***

Never say die! Sometimes, you will find that you are as close to winning as you can be, even though things have never looked worse. This is especially true in gold mining.

This is not to say that you should keep pushing hard in a direction that is not working. It is to say that if one direction does not seem to work, take it a little further, and then look around for another way to go. Try to avoid getting fixed in one mind-set. Ask yourself what you are trying to accomplish, and keep your awareness open and your imagination working to find the available opportunities to accomplish your goal.

We manage group gold mining operations in Northern California just about every other weekend during the spring, summer and fall months. During these operations, it is my personal responsibility to make sure that silt, sand and gravel are not washed back into the river off the bank from our tailings-water. This requires continuous effort to keep tailings-water directed along the natural contours on the land outside of the river. Sometimes I spend a lot of time running up and down the bank trying to stop the water or slow it down. One thing for sure; if you stop the water in one place, pretty soon it is running down the bank somewhere else! Gravity is a force of energy that never stops.

In many ways, this is not dissimilar to a person's intention to accomplish a goal. Sometimes we will find our immediate progress slowed down or stopped altogether. But if we maintain our intention and study the barriers to our progress, we can usually find other ways to make progress. The key is in maintaining the intention, keep pushing along and having a little patience.

The best things in life usually take a little time to accomplish. And the best memories often come from the accomplishment of difficult goals.

Having watched hundreds upon hundreds of dredging operations over the years, if I had to pull together my best advice in a single statement, it would be: ***Hope is what got you started. And your own hope is what will carry you through to a successful finish. So don't be too quick to give it up!***

You can do it!

*Dave McCracken*

# 1

# PAY-STREAKS – WHAT THEY ARE

## IT ALL HAPPENS DURING MAJOR STORMS

It only takes a small amount of water force to move sand-sized particles of rock downstream in a waterway. It takes more water force to move pea-sized gravel down river, and even more force to move golf ball-sized rocks. The larger the size of the rock, which generally all consists of the same type of material, the more force of water that it takes to wash it downstream in a waterway.

The shapes of rocks also play an important part in this. For example, a round rock will take less water force to move than a flat rock which has the same amount of mass. This is because the round rock has more surface area for the water to push against, and a rounder-shaped rock is more prone to roll along than something which lies flat on the bottom.

It also takes a greater amount of water force to move larger-sized particles of gold, than it does to move smaller particles.

Some gold-bearing waterways contain a substantial volume of gold which is so small in particle size that it exists in a state of suspension within the river water. Some amount of gold is moving downstream in any gold-bearing river at any given time.

An increase in water flow will also increase the amount and size of streambed material and gold being moved downstream.

Generally, the amount of gold being moved downstream within a waterway during periods of non-flooding is insignificant to a gold dredger. This is because suspended gold is so small in size that it defies recovery methods used by most suction dredges. Suspended gold is also so widely dispersed in small amounts throughout the river, that it simply is not worth going after in itself with a suction gold dredge.

While average winter storms may move substantial volumes of fine and ultra-fine gold down gold-bearing waterways, this gold is likely to be so widely dispersed throughout the overall river, that it is also not of great value to the gold dredger.

A gold dredger is mostly interested in that gold which lies inside and underneath of naturally-formed hard-packed streambeds. For the most part, this gold will stay right where it is until a storm of major proportions comes along and causes so much water force in the waterway, that large sections of formed streambeds, and the gold that is with them, are swept up and washed downstream within the waterway.

When a storm or snow runoff of very large dimensions comes along and creates so much water force that large portions of the riverbed are torn up and washed downstream in the waterway, large portions of the bedrock will also get pounded loose. Gold which was trapped in that bedrock will also get washed further downstream along with the rest of the streambed material.

There is a lot of gold remaining in today's gold-bearing waterways. Some was left behind in low-grade deposits which the early-timers were not interested in mining. While there might have been a lot of gold in some sections of riverbed, it might also have been too widely dispersed or sitting underneath too much valueless streambed material to make it worth mining in those days. Some very rich deposits were entirely missed because they were out of sight. Without processing every bit of streambed, which was overwhelmingly impossible, the early-timers simply could not find all of the rich deposits.

A lot of gold was washed out of the high gravel deposits by the early-timers through sluice boxes and right back into the present rivers and creeks. This was particularly true of hydraulic mining, where an estimated 50% of the gold was missed by many of the large and small operations alike. Hydraulic mining used large volumes of water under great pressure. Efficiency of gold recovery was often very poor.

The last 150 years of natural and man-caused erosion has washed gold into the present rivers and streams out of the higher and older dry streambeds, and perhaps even out of some lodes that are still in existence. Although lode gold is probably not as large a contributing factor as these other sources of new gold into today's waterways.

In taking all of this into consideration, we are talking about a lot of gold still being present in today's gold-bearing rivers.

Here is the important thing to know: ***Almost all of the gold deposits we are looking for in active waterways were created during major flood storms.*** When a major storm occurs in a gold-bearing area and tears up large portions of streambed or the tailings from previous mining operations, the gold which existed there is set free and put into motion downriver. Because if its superior weight (six times heavier by volume than the average of rocks, sand and silt within the waterway), most of the gold will travel along its own path within the waterway and will be deposited in common areas along the very same path. This is the gold that the modern gold dredger should be searching for.

The same major flood storm which causes enough force to tear up large portions of streambed material and wash it downstream will also redeposit much of that material into newly-formed, hard-packed natural streambeds in the same areas which were successfully mined before by the early-timers. ***It does not matter if the area had been mined before. A major flood storm creates a new beginning in that area.***

Because gold is so much heavier than average streambed material, most gold quickly works its way to the bottom of the streambed material that is in movement during a major flood storm. Sometimes it is hard for beginners in this field to understand this.

This is because when we are out dredging during the summer months, the streambed consists of rocks, boulders, sand and gravel which is all compacted together into a solid structure. It is hard to visualize gold working its way down through all this hard-packed material. However, it is important to realize that the streambed was formed at a time when all the material was a fluid slurry, similar to the way gravel washes through the sluice box in a dredge. When the streambed material was in fluid movement along the bottom of the waterway, the six-fold downward pressure of gravity upon gold drove most of it quickly to the bottom of the material in movement. Under these circumstances, the gold deposits formed before (and underneath) the streambed material was formed into the hard-packed beds that we see in the river.

If we placed 10 pieces of fresh Styrofoam in the moving water of a river, few people would doubt that all ten pieces of Styrofoam would quickly float to the water's surface. This is because everyone knows that Styrofoam is so much lighter than the water. The very same rule applies to gold. If you put 10 flakes of gold into streambed material that is in movement during a major flood storm (or washing through a sluice box), you are just as likely (as in the case of the floating Styrofoam) to find all 10 flakes of gold deposited along the lower surface of the streambed material that was in movement. This leads us to one of the most important fundamentals in dredge prospecting: ***Most high-grade gold deposits are located along the bottom of a hard-packed streambed layer.***

Since most high-grade gold deposits will be located along the bottom of hard-packed flood layers, your sampling program should target the bottom of streambed layers.

For the most part, normal winter storms occurring in gold country do not generate enough water force and turbulence to redistribute hard-packed streambed layers. A winter storm might be enough to sweep up small portions of streambed in faster-moving sections of the river and redeposit new streambeds in those areas. But this small amount of movement is not likely to put paying quantities of gold into movement within the waterway. It takes a substantial amount of streambed in motion to place substantial amounts of gold into movement. This causes a whole river to form new placer deposits. Such storms occur occasionally, and are the main cause for a streambed cutting deeper into the earth over geologic time. Most gold-bearing areas have had at least one or two of these major flood storms since the gold rush days of the mid-to-late 1800's.

After a major flood storm has been through an area, the plant growth and underbrush which normally grows on the gravel bars will be gone. This will also be true with most of the growth, and even many of the trees, along the riverbanks. If such growth remains after a storm, it means the streambeds which existed prior to the flood still remain in place. Since trees and other vegetation are anchored into pre-existing streambed structures along the edge of waterways, if the streambeds are swept up and washed away, there will be little or no vegetation remaining in those places.

It takes an incredible amount of water force to cause an entire riverbed to get washed up and moved downstream, but this is what it takes to form many new placer gold deposits within the river.

Gold which is washed down along the bottom of fluid streambed material during a major flood storm will eventually get trapped along the stationary surface that it is being washed across. This might be bedrock, or it might be the surface of a pre-existing layer of streambed that remains in place during the storm.

## DIFFERENT KINDS OF STREAMBED PLACER DEPOSITS

There are several different kinds of streambed gold deposits which a gold dredger should be aware of, because each needs to be searched for and mined in a different way – and each has a different meaning when we find them during sampling.

***TAILRACE & BENCH BLEED-OFFS:*** A high percentage of the gold in today's rivers was once washed into them out of the higher and older streambeds resting up alongside the rivers, or from others which are located some distance away. This is the result of millions of years of massive flood storms and natural erosion. Today's storms continue to wash gold out of the banks and into the present rivers and creeks. That gold which was set free by earlier mining activities, but missed by their recovery methods, is also being actively washed into the present streams and creeks.

Because gold is so heavy, it tends to follow and deposit along a common path within the waterway during a major flood storm. Most pay-streaks within the waterway will be found along this common path.

But when a new source of gold enters the river from a point off this common path, it may need to wash some distance downstream before it finally arrives in the common path which other gold is traveling in the waterway. Therefore, these new deposits of gold eroding from high gravel deposits and tailraces from

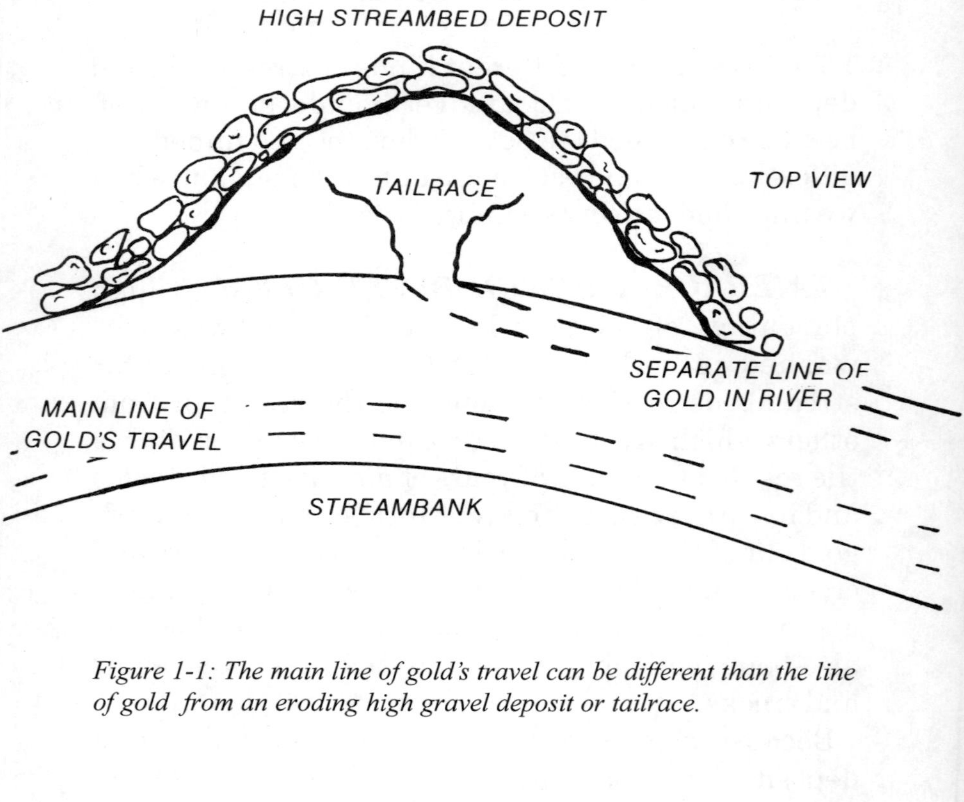


*Figure 1-1: The main line of gold's travel can be different than the line of gold from an eroding high gravel deposit or tailrace.*

earlier mining operations are likely to follow their own path downriver until they join up with the main path.

So you would need to look for this type of deposit usually in a different section of the river than you would normally look for pay-streaks. (See figure 1-1).

***PAY-STREAKS:*** Pay-streaks (also called *"stringers")* are the most common, and best paying, and generally easiest to locate, type of placer gold deposit being found in most of today's gold-bearing rivers. A *"Pay-streak"* is a concentration of gold which has formed in a section of river because of a sudden or gradual lessening of water velocity in that area during a major flood storm.

During an extreme high water period, when a whole layer of streambed material is in fluid movement along the bottom of the river or creek, because there are many sudden and gradual changes in the river's direction and downward slope, different areas along the bottom of the waterway will have greater or lesser forces than others. Therefore, different push/pulls are set up along the river's course. There will be high-pressure areas, medium-pressure areas, low-pressure areas, and even back-pressure areas. When you see water moving backwards against the main flow of water, as in a back-eddy, you are looking at a back-pressure area. Where you see the water slow down or even stop, in comparison to the main flow of water, you are looking at a low-pressure area. Many low-pressure and back-pressure areas are created along the river's path during a major flood storm. These are the places where pay-streaks form, as long as they are located along the path which gold is following as it is washed downriver.

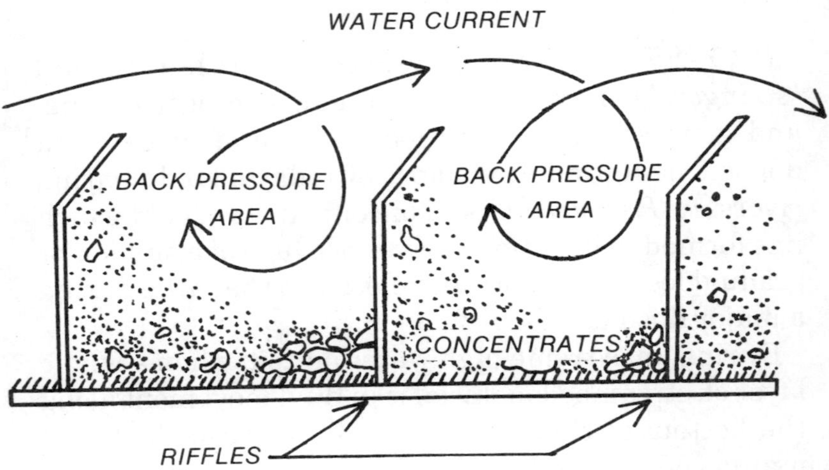


*Figure 1-2: A vacuum is created just behind a Hungarian riffle when water is passed over it.*

A standard Hungarian-type riffle within a sluice box is an ideal example of a back-pressure area. Behind the riffle, a vacuum (back-pressure) is created when a flow of water is run over the riffle. The vacuum will suck water and material into the riffle as they pass over. But because the riffle will only hold so much material, lighter materials are forced out of the riffle, while the heaviest materials (gold and iron) will remain. This is because there is not enough force behind the riffle to push the heaviest material out again. (See figure 1-2).

Similar conditions to this will occur along a waterway during a major flood storm, setting up back-pressure areas like this on a much larger scale. Many of the common types of low/back-pressure areas that are found in a river are outlined in my book, ***Gold Mining in the 21st Century***.

Low and back-pressure areas can also be formed because of some major change off to one side of the river, or when the river makes a change of direction. One good example of this is the common low/back-pressure area that will be found at the tail end of most bends which a river takes. If you take a good look at where a river changes direction, you will see that the flow of water and material will come on in full force at

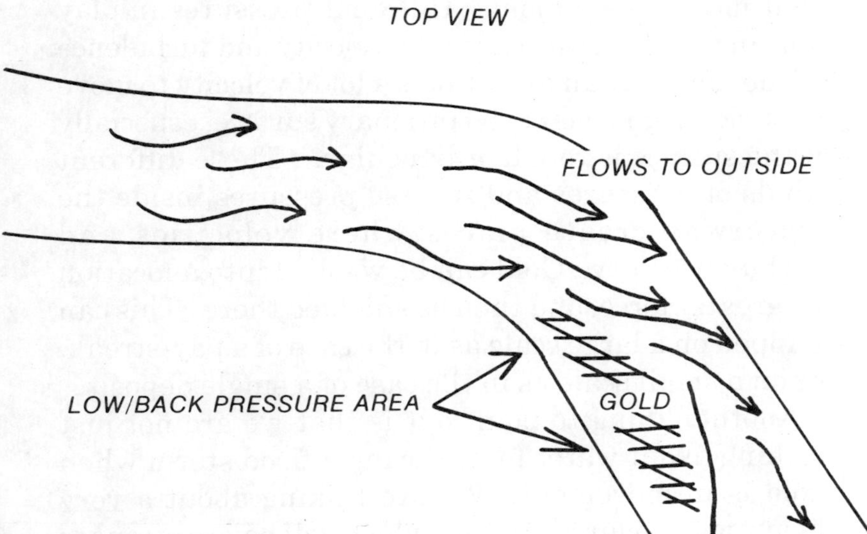


*Figure 1-3: How a low/back-pressure deposit is created at the inside of a river bend.*

the head of the bend. Since centrifugal force will cause a greater portion of the force to flow to the outside of the bend, there will be a lowering of pressure, or even a reverse of pressure (back-pressure), to the inside of the tail-end of the bend. And sure enough, the inside tail-end of a river bend is a very common place to find a sizeable pay-streak in gold-bearing river.

Therefore, pay-streaks can be caused by a change in pressures off to one side of the river, or by a change in pressures due to the difference in downward slope of the river's bedrock foundation. (See figure 1-3)

It is important to point out here that it really is not the changes in pressures that cause gold to deposit. It is the changes in water velocity and turbulence. It is the friction of water's flow across the surface of objects that moves them along. Different pressures in play within the waterway cause the velocity and turbulence of the flow to change. It takes a lot of velocity to move particles of gold across a stationary surface, especially if the surface is rough or irregular. These different kinds of pressures and reverse pressures inside the waterway create places where velocities and turbulence vary. Gold can be washed into a location by greater force, and then be shielded there. This can happen on a large scale as in the case of a pay-streak, or on a small scale as in the case of a single deposit.

Another thing to point out is that we are not just talking about water force during a flood storm when gold is in movement. We are talking about a very heavy water-slurry mixture which will have a stronger force to move the gold along.

***SINGLE DEPOSITS:*** A single deposit of gold can form anywhere along the general path in the river that gold follows, where an obstruction, a small change in the bedrock or a crevice is located which creates a dead spot from the main force of velocity and turbulence during a major flood storm. Here, you are going to find a single deposit inside of the crevice or behind the small obstruction or other small hide-away from the main flow. Sometimes a single deposit will be found that carries a substantial amount of gold. Single deposits like this will only be found along the general path that gold travels along downriver, or along the path gold is following when eroding from some new source of gold into the waterway – like a tailrace or a high gravel deposit. (See figure 1-4.)

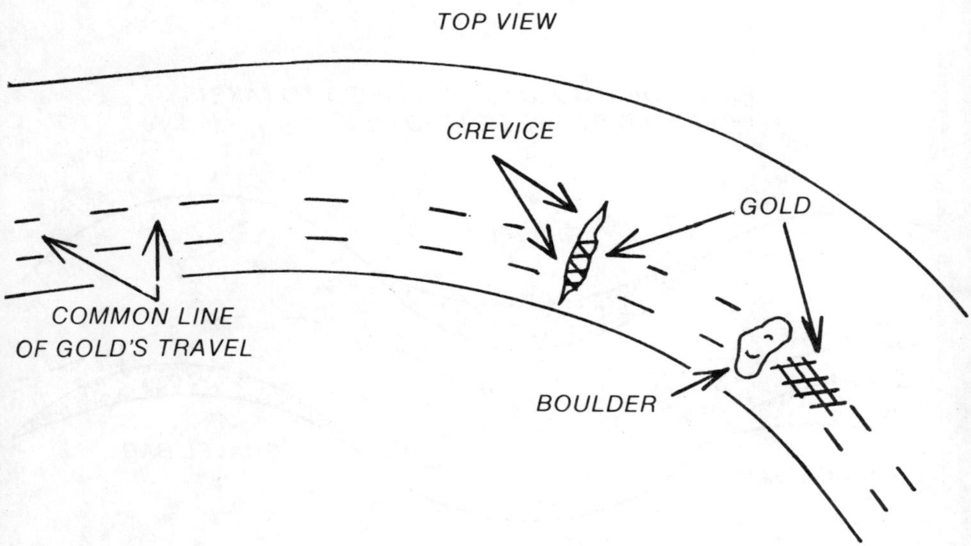


*Figure 1-4: Single gold deposits will be found along the common line of gold's travel.*

Because a single gold deposit is confined to one small area, and all signs of its presence are usually well hidden by streambed material which lies over top, single deposits are generally more difficult to find than pay-streaks. In the case of pay-streaks, many of the major and minor changes in the river's flow can be directly observed, and the location of their resulting low and back-pressure areas can be predicted to exist within the vicinity. Pay-streaks cover some width and distance and tend to be easier to locate by sampling. Because pay-streaks usually carry a sizeable amount of gold, you will find that most experienced gold dredgers are going after them, rather than single gold deposits.

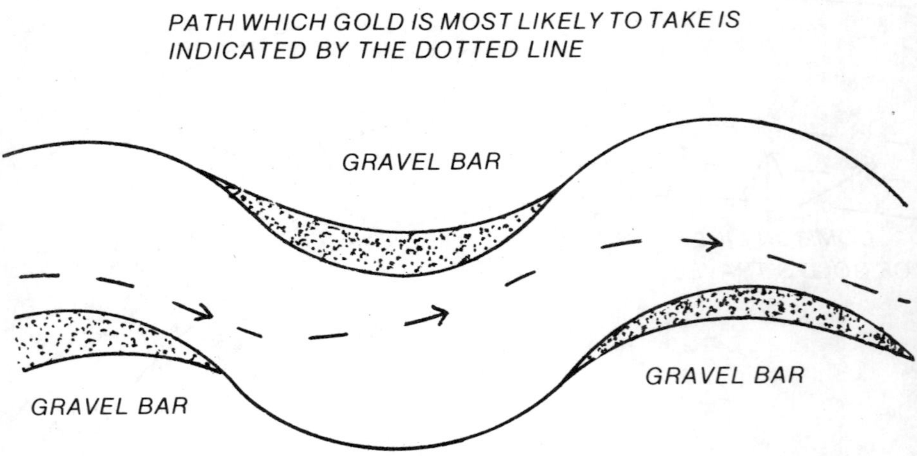


*Figure 1-5: The general path which gold tends to deposit along.*

## THE GENERAL PATH OF GOLD

Because gold is around six times heavier than the average of other streambed materials that get swept downriver during a major flood storm, most gold tends to follow and be deposited along a common path within the waterway. For the most part, this path seems to follow to the inside of the bends, and the shortest route between the bends, as shown in figure 1-5.

There are so many different forces at work inside of a waterway during a major storm, and there is so much material being tumbled downstream, some of the gold does not travel down along this path. But most of the gold does deposit along a common path in the waterway. Most of the gold we recover out of high-grade single and pay-streak deposits with suction dredges will be found along a common path. Historical statistics show that a far greater percentage of single deposits and pay-streaks have been found along a common path, than have been found elsewhere within the waterway. So it is wise for you as a gold dredger to know about the general path, and to direct your sampling activities towards it.

The general line of gold's travel is largely affected by the various major low/back-pressure changes within the stream's flow, especially those that pull from off to one side. Remember this: ***The general path that gold follows is usually most definite just after a major directional change in the river, where the substantial low/back-pressure to the inside pulls the line into a concentrated zone.***

On a long stretch of straight river, where there are no major low/back-pressure areas off to either side,

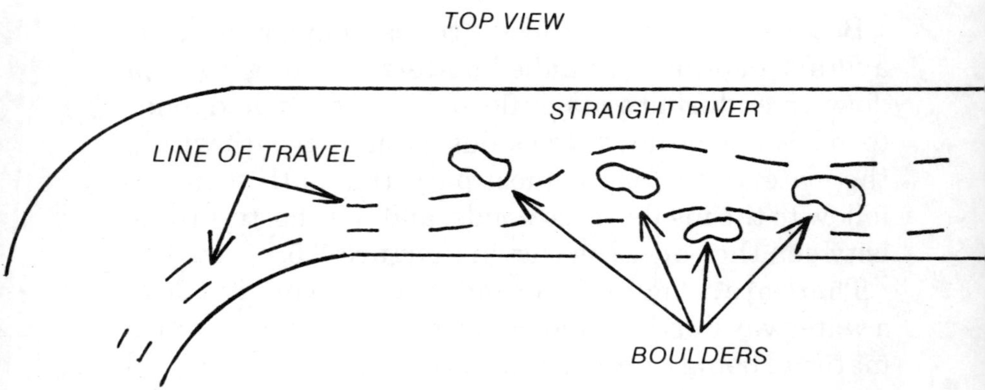


*Figure 1-6: The general line of gold's travel can cover a less-defined area along a long, straight stretch of river.*

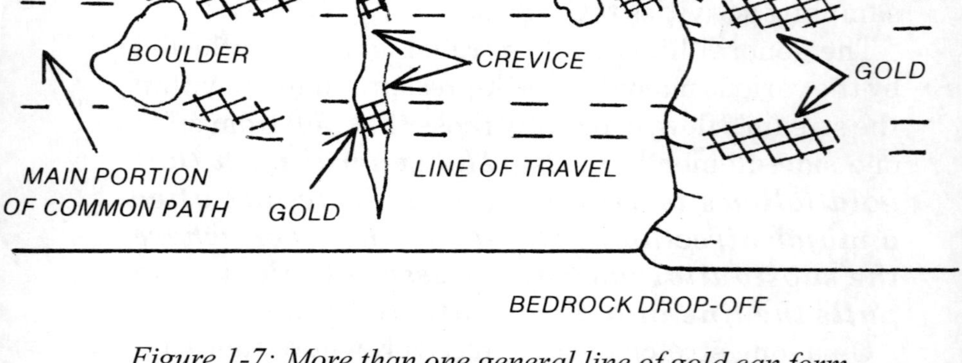


*Figure 1-7: More than one general line of gold can form along a straight stretch of river.*

the minor ones tend to disperse the general line of gold out into a less-defined area of travel, as shown in figure 1-6.

Sometimes, on a long stretch of straight river, small and medium-sized low/back-pressures or obstructions can cause more than one general line of gold to form down the length of river. In this case, single deposits and pay-streaks can be found and recovered along each separate line, as shown in figure 1-7.

Sometimes, there can be more than one path of gold down a river even when there is a directional change in the waterway. These additional lines are usually the result of some new source of gold eroding into the waterway from upstream. Or they might also result from gold being widely dispersed throughout the river because of a long stretch of straight river upstream, as shown in figure 1-8.

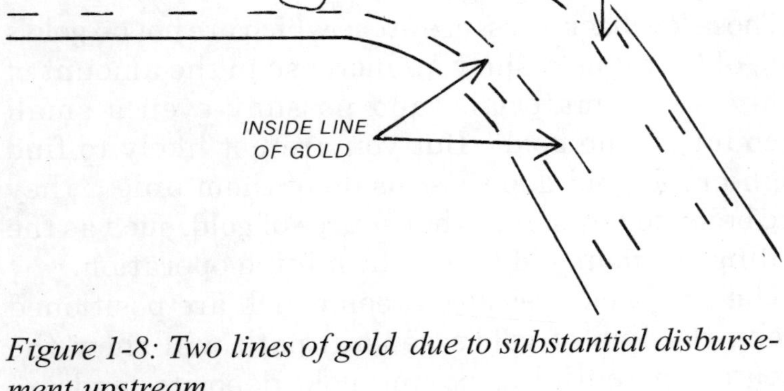


*Figure 1-8: Two lines of gold due to substantial disbursement upstream.*

Other heavy minerals and materials within a riverbed also concentrate along and follow this same general line that gold travels. Iron rocks, black sands, old nails, pieces of lead, metallic tools and other heavy man-made objects commonly follow the same path as gold down a waterway during a major flood storm. However, because iron and lead do not have as high a specific gravity as gold does, the path that these follow can be a wider one and have less-defined boundaries than the line which gold travels.

This is also true of the fine and ultra-fine-sized particles of gold. The smaller that a particle of gold is, the less it is likely to be affected by the various low/back-pressure areas within a waterway. The greatest concentration of fine gold and other heavy materials, though, does often follow the same general line of travel that larger particles of gold normally deposit along.

Please keep in mind that low and back-pressure areas are created all throughout a waterway during a major flood storm. It is the low and back-pressures which exist along the general gold path that we are looking for in dredging.

Those low/back-pressure areas which are not on gold's line of travel may show an increase in the amount of other heavy materials, and possibly even a small amount of fine gold. But you are not likely to find high-grade gold deposits inside of them unless they are being fed by some other source of gold, such as the tailings from an old hydraulic mining operation.

The low/back-pressure areas which are positioned along the line of gold's general path will show the greatest potential of paying gold deposits, and are where you often also find the greatest amount of other heavy materials too.

Sometimes you can find the common gold path by sampling around to find where in the river that the greatest amounts of heavy materials are concentrating. From there, it is possible to sample forward along that line to find where the low/back-pressure areas (pay-streaks) are located. If there is an increase in the heavier minerals and materials showing, and if they are along the common gold path, there should also be at least a slight increase showing of gold along that line. This is a good sign that you are on the right path to find the paying deposits (See figure 1-9).

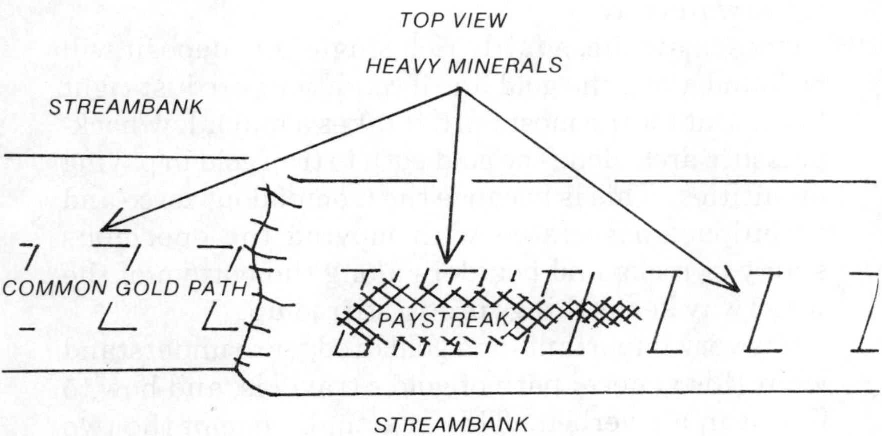


*Figure 1-9: An increase in gold and other heavy minerals and materials in the streambed is a good sign that you are on the common gold path.*

Gold does not deposit in paying quantities everywhere along the general line of travel. Although, you are more likely to find small traces of it having passed along the gold line in places here and there where small amounts were able to become trapped (very small single deposits). Very rough and irregular bedrock along the general line will often show traces of gold, whereas smooth and polished bedrock along the general path often will not.

Think about it like this: During a major flood storm, most of the gold is being washed down across the surface of bedrock or a layer of pre-existing streambed that remains in place during the storm. Gold is extremely heavy, so it leaves traces along its path. If 90 percent of the gold washes down along a fixed path in the river, you will find substantially more traces along the path, than off of the path, in most portions of the waterway.

Once in awhile, a fairly rich single gold deposit will be found along the gold line if conditions are just right for it. But for the most part, it takes a major low/back-pressure area along the gold path to trap gold in paying quantities. This is because the tremendous force and turbulence associated with moving the enormous slurry of rocks and boulders along the bottom of the waterway keeps things pretty stirred up.

It is very important for a gold dredger to understand what this general path of gold's travel is, and how to find it in a riverbed. This is actually one of the two primary fundamentals which guide a good sampling plan:

(1) Gold follows its own path in the waterway, and;

(2) Gold deposits along the bottom of hard-packed streambeds.

The first step in a dredge sampling plan is to locate the gold path within the waterway. The second step is to establish which storm layer(s) in the streambed that the gold is traveling along (at the bottom of). The third step is to locate the pay-streaks at the bottom of that layer(s), along that path. This is where almost all of the high-grade gold will be found in the waterway.

Locating the gold path can be accomplished by dredging a series of sample holes across the width of the waterway and comparing how much gold is recovered out of each sample hole. At this stage, you are looking for the path where the most gold seems to be traveling down the waterway, and at which layer within the streambed most of the gold is traveling.

You can always depend upon at least one path being present, as long as gold has traveled through that section of river during a major flood storm. The exception to this is when the gold has been thoroughly dispersed by a long straight stretch of river.

So let's summarize here: The gold we are looking for lies at the bottom of hard-packed streambed layers which were formed during major flood events. Because gold is so much heavier than the average of other streambed material (rocks, sand and gravel) which washes downstream during a major flood, the gold concentrates along a narrower path of its own in the waterway, and moves downstream across the stationary surface of bedrock or a hard-packed layer of pre-existing streambed from some earlier storm.

Gold deposits along this surface either by leaving traces, by dropping into single gold traps, or by forming sizable pay-streaks where water velocity and turbulence lets up on a large scale due to changes or obstructions in the flow.

## THE CHARACTERISTICS OF PAY-STREAKS

In earlier books on the subject of gold mining, mention is made that gold will deposit in the back eddies. This is true - to a point. First of all, for large amounts of gold to collect in a back eddy (back-pressure area), it will have to be located along the path that gold follows in its route downriver during a major flood storm.

And second, if too much force of water is moved over the top of a back-pressure zone, it can become a boil-out area (high velocity and turbulence) in which very little concentration of gold values can take place. An example of this is in the case of a Hungarian-type rifle. When too much water velocity is flowed over a Hungarian riffle, like when the slope of the sluice box is increased too much, the force of the flow being pulled into the vacuum behind the riffle becomes so turbulent that even most of the gold can be washed out of the riffle. The force within the back-pressure area behind the riffle can be increased to the point where no concentration at all will take place. (See figure 1-10.)

A back-pressure area within a waterway can also become a boil-out if too much water force is passed over it during a major flood storm. This is one of the reasons why you may not find the amount of gold that

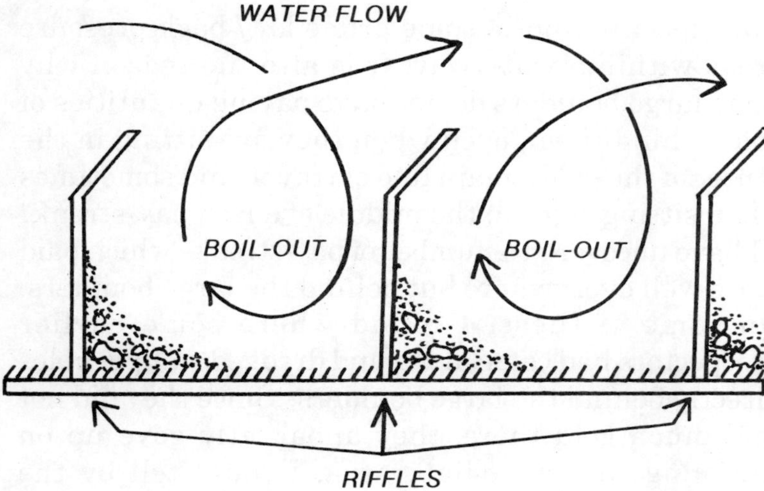


*Figure 1-10: Too much water force over a back-pressure area can cause a boil-out.*

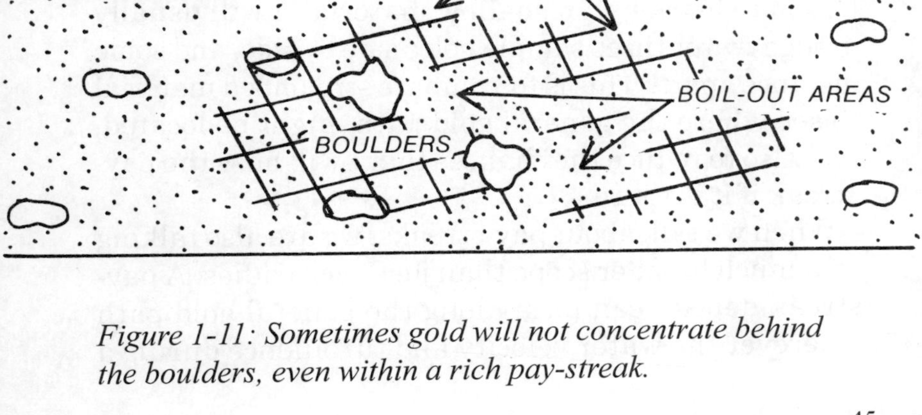


*Figure 1-11: Sometimes gold will not concentrate behind the boulders, even within a rich pay-streak.*

you expect to find in some prime low/ back-pressure areas within a waterway. It is also the reason why some large boulders do not have paying quantities of gold behind them, even when they are sitting in the center of the gold's main line of travel, and sometimes when sitting right in the middle of a rich pay-streak.

I have uncovered a number of pay-streaks which paid quite well everywhere but behind the large boulders. On some of these, I found where some earlier prospectors had come along and dredged sample holes directly behind the large boulders. Since they did not find much gold there, they apparently gave up on sampling the immediate area. I could tell by the streambed formation that, other than where the sample holes were dredged behind some of the boulders, no other mining had been done within the vicinity of the pay-streaks. The gold simply did not deposit behind them. Yet they were sitting right smack in the middle of rich pay-streaks. (See figure 1-11.)

Had those prospectors moved their holes just a little to the outside of the boulders in either direction, they would have been into a rich gold deposit. This is a good lesson to learn by. In shallow material, I will usually dredge two or three sample holes side by side, and some distance apart. This is to compare streambed material to see where the general gold path might be located, and also to reduce the chances that I will miss the pay-streak if it is there.

When we talk about pay-streaks, we are also talking on a much broader scope than just back eddies. A pay-streak deposit can form along the general gold path wherever the water velocity and turbulence during a major flood storm lets up enough to drop gold, but still moves fast enough to sweep off the lighter materials. I have recovered some rich pay-streaks right off of smooth and polished bedrock, where the flow let up just right during a flood storm to drop gold. While this happens, places where there is a lot of rough and irregular bedrock inside of a low/back-pressure area are more likely to provide a good pay-streak. That, or where the pay-streak lays down on top of another pre-existing layer of streambed that remained in place during the flood storm.

There can also be low/back-pressure areas within low/back-pressure areas. Where there is rough and irregular bedrock in these places, you are likely to find the richest-paying deposits. For example, I know of a place where a medium-sized gold-bearing river travels for several miles through a rather narrow gorge with a steady downward grade to it. Then, suddenly the riverbed widens out to a large degree and the downward grade lessens up for a few miles. So there is pretty fast water for a few miles, and then a few miles of slower water. The entire section of slower river is a long low-pressure area, compared to the higher pressure section of river up ahead of it, as shown in figure 1-12.

That section of slower water in this example also has bends, short sections of faster and slower water, outcroppings, upcroppings, obstructions and major changes in bedrock which set up smaller major low/back-pressure areas inside of the long length of slower moving water. For its entire length along this section of river, each and every obvious low/back-pressure area has carried a substantial pay-streak. There are

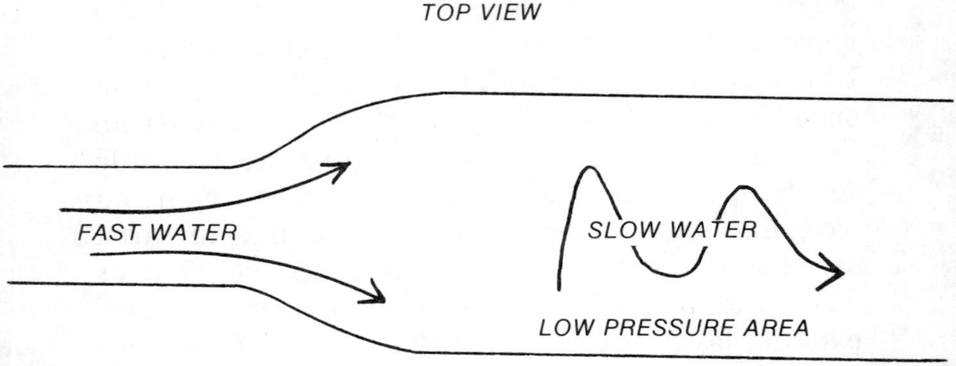


*Figure 1-12: Fast water (high velocity) section of river flows into a slow water (lower-pressure) section.*

also several long straight stretches of river which resulted in at least three separate and well-defined lines of gold in some areas. Because the bedrock was right for it, and because the whole stretch was one huge low-pressure area, there were pay-streaks to be found along each line of travel wherever any low/back-pressure area was set up in the stream's flow. Indeed, in many places it was a paying proposition to just get onto any of the gold lines and start dredging.

One interesting point about this is the entire riverbed throughout that long slow section of river was thoroughly mined during the late 1800's. It paid very well then, too! More recent major flood storms have redeposited the area again with fresh gold for modern-day gold dredgers!

So you can find low/back-pressures within larger-sized low-pressure areas, and this is one thing you'll want to keep your eyes open for.

In some situations, it might take some distance for the water force to slow down enough in a long low-pressure area to start dropping gold in large amounts. Where the river widens for a long distance after coming out of a long narrow passage, the force of water might travel some distance further before the velocity lets up enough to allow the major low/back-pressures to concentrate gold instead of boiling it out. One indication to look for, which can be a big help, is where the material starts to collect in large amounts along the riverbed, as in large river bars, boulder-packs or bedrock islands sticking out of the waterway. There, and perhaps for some short distance upstream, is where the pay-streaks ought to have formed. Your sampling efforts will uncover the story. (See figure 1-13.)

## HOW THE SIZE OF A WATERWAY CAN AFFECT PAY-STREAKS

Generally, the larger in size that a waterway is, the more pay-streaks you are likely to find along its course, and the better formed they are likely to be. How much gold they will contain will be largely affected by how much gold washed down through the different sections of river during the major flood storms which created the streambeds that exist there now.

Because a river is so much larger than a small creek, in comparison to the size of material (boulders) that is swept downstream during flood storms, there will be far fewer major obstructions which seriously affect the

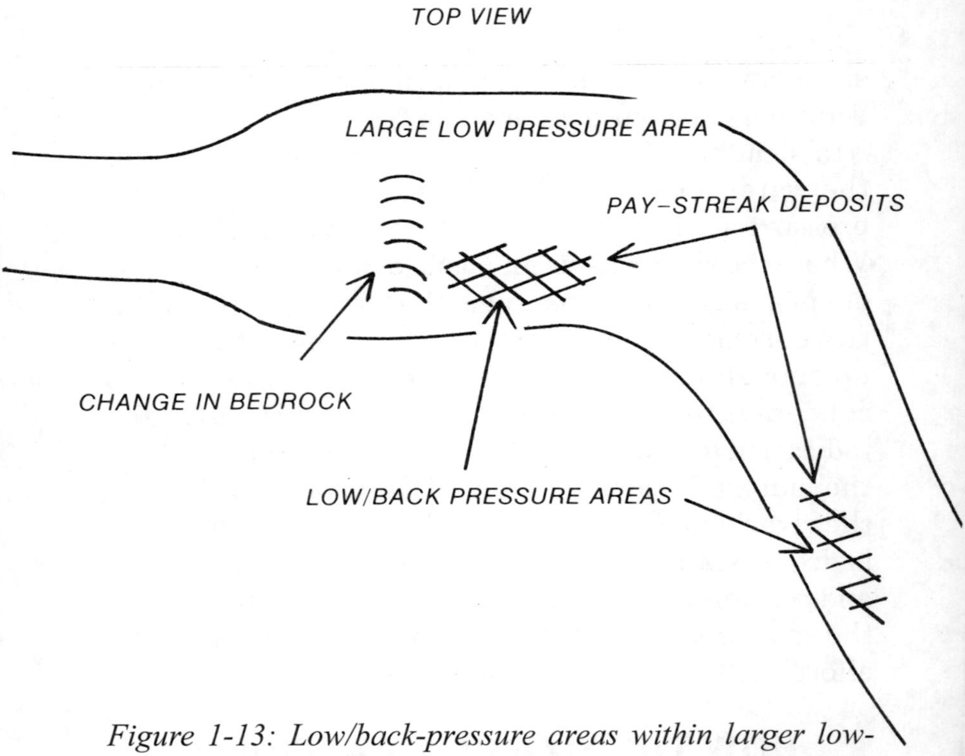


*Figure 1-13: Low/back-pressure areas within larger low-pressure areas are excellent places to find concentrated gold deposits.*

flow of water and material. So the general line of gold' s travel, and those areas of concentration, are likely to be more orderly and well-defined. In addition, the low/back-pressure areas in a large body of running water are going to be of far greater size than those found in a small body of moving water. So the pay-streaks are going to be larger, too.

You are <u>not</u> likely to find as many pay-streaks in a smaller-sized waterway. Instead, it is more likely that you will find single deposits. The richer that the stream is, the more deposits you will find in a small stream. Smaller streams, unless they are very rich, tend to be rather spotty with deposits. You will find gold for awhile. Then it will just quit. Then you might pick it up again.

Richer rivers also carry more and richer pay-streaks. So the larger the size of the waterway that you dredge, the fewer deposits that you will need to locate; because you will spend more time dredging them up and less time looking for them. I know of many pay-streaks that went on for a number of years before playing out. Different sections of gold-bearing rivers will be found to contain more and richer pay-streaks than other sections.

The other side of this is that some larger rivers have deeper streambeds along the bottom. This can dramatically increase the amount of work required to complete each sample hole. So finding the pay-streaks in larger, deeper rivers can be more difficult to accomplish. But this is not always the case. Each river has its own unique set of circumstances.

## THE BOUNDRIES OF A PAY-STREAK

A pay-streak can be of almost any width, even as wide as the whole river.

In some pay-streaks, the outside boundaries are less defined and visible than in others. Sometimes the outside edge of a pay-streak can end so abruptly that if a sample hole is dredged just a few feet to the outside of it, you might not see any sign of its presence. (See figure 1-14.)

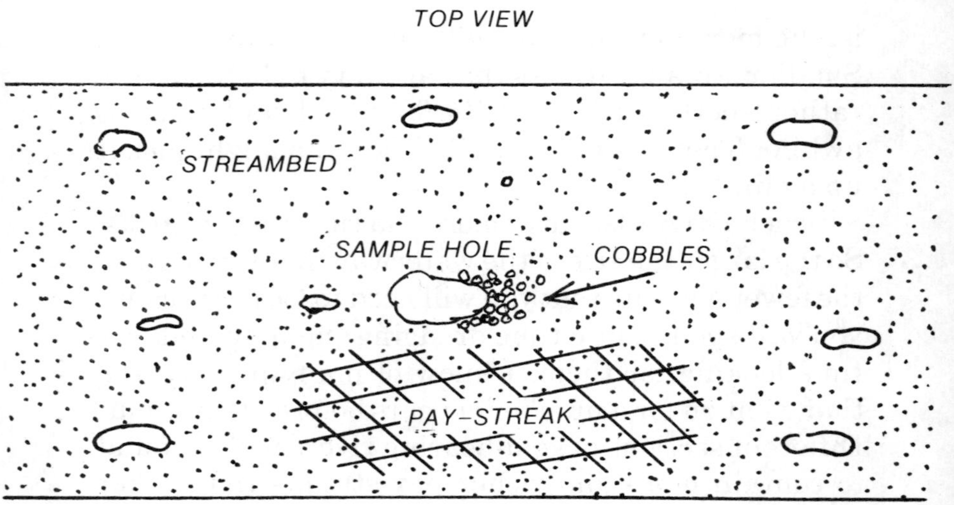


*Figure 1-14: Some pay-streaks have outside boundaries which end abruptly.*

In some pay-streaks, the concentration of gold will grow gradually less as you move outward from the richest portion of the pay-streak. In this case, a sample hole dredged slightly to the outside of the main portion of the pay-streak is still likely to show some sign of its presence (See figure 1-15).

Sometimes a pay-streak will pay rather consistently all throughout. Other pay-streaks will be somewhat spotty, paying good for awhile, then poor, and then good again. Sometimes the bedrock along a pay-streak will change from rough to smooth, and then rough again, with rich gold concentrations only lying on the rough sections of bedrock (See figure 1-16).

Pay-streaks are most likely to form either on the bedrock or in the contact zone between two layers. Because gold is so heavy, it is unusual to find high-

TOP VIEW

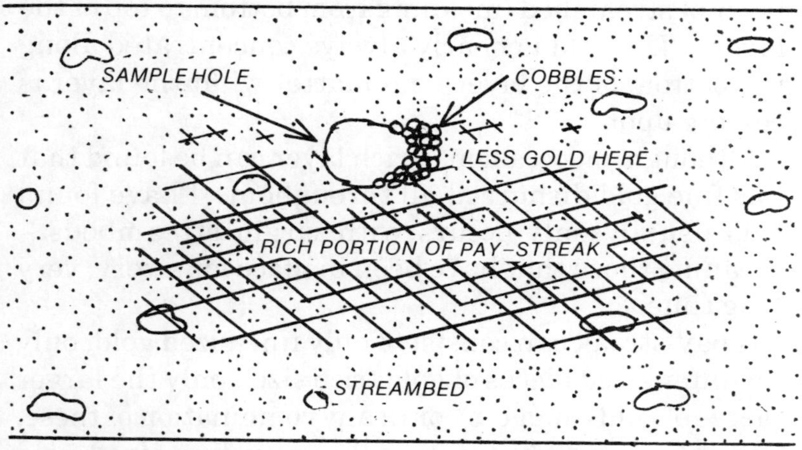


*Figure 1-15: The amount of gold tends to grow less further out from the richest portion of some pay-streaks*

TOP VIEW

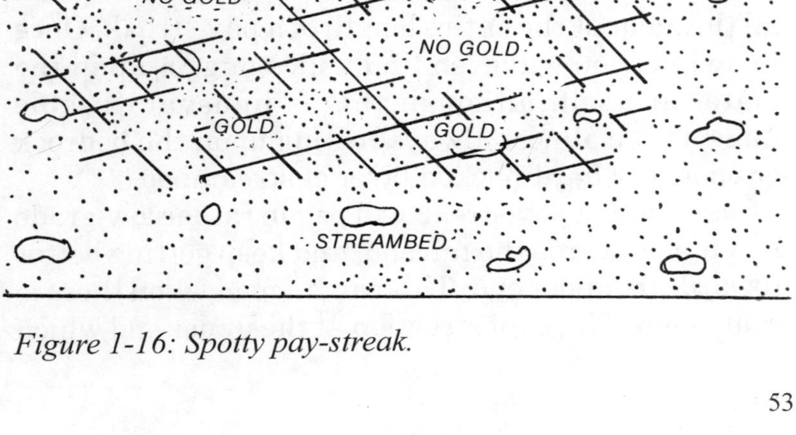


*Figure 1-16: Spotty pay-streak.*

grade deposits with the gold disbursed throughout a layer of streambed (meaning from bottom to top of the layer). The gold is nearly always concentrated along the bottom of the layer – no-matter what the layer is resting upon.

Although, on occasion, a rich layer can be found that has fine gold disbursed all throughout. I have found this usually only occurs with virgin streambeds – meaning those that have been in place for a very, very long time.

A pay-streak can consist of only fine-sized gold, only medium-sized pieces of gold (rice size), only the larger pieces of gold (nuggets), or in any combination of these. Sometimes a pay-streak will pay in fine-sized gold towards the tail end, medium-sized gold in the center, and in nuggets towards the upper-end of the pay-streak. Or, because different-sized particles of gold have different resistances to water force, in some pay-streaks you will find the larger pieces hung up on a certain line along the pay-streak's length. Once in awhile, you will find all the big gold just to the outside of what appears to be the outer boundary of the richest part of the pay-streak (See figure 1-17).

Sometimes all the different sizes of gold will be found all throughout the entire length of a pay-streak. Once in awhile, you will even find a nice nugget up in the material itself, although this is unusual. Sizable nuggets are almost always resting upon the bedrock or on a very hard-packed layer of streambed.

Sometimes a pay-streak will pay in rather low-grade amounts towards the tail end, and keep getting better towards its upper end. This can happen when there is some kind of large obstruction at the upper end which

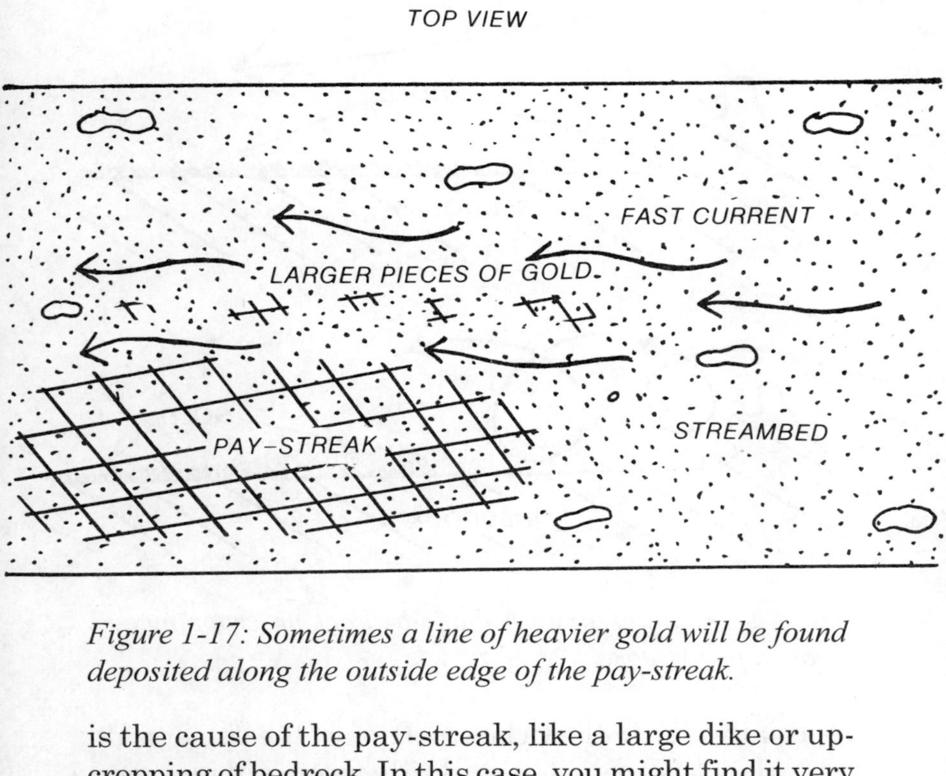


*Figure 1-17: Sometimes a line of heavier gold will be found deposited along the outside edge of the pay-streak.*

is the cause of the pay-streak, like a large dike or up-cropping of bedrock. In this case, you might find it very rich just behind and around the obstruction. Sometimes that area might be entirely boiled out. Each case can be different, as shown in figure 1-18.

Sometimes a pay-streak will pay rather consistently all throughout, but become spotty towards its outside edges. This can make it more difficult for a dredger to follow the outside boundaries and recover all the paying-quantities of gold without also having to dredge up a bunch of low-grade material, too (See figure 1-19).

As mentioned earlier, pay-streaks also contain an increase in iron rocks and black sand. This is

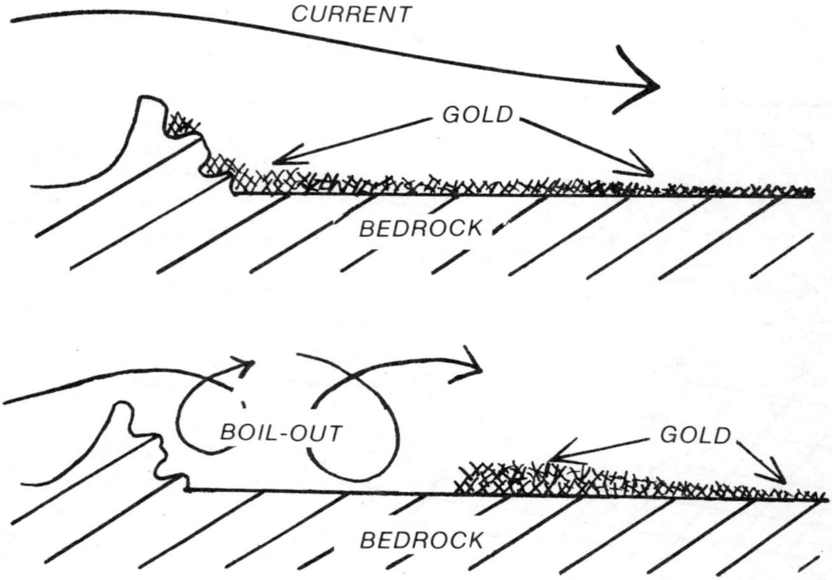


*Figure 1-18: Different ways that a bedrock up-cropping can create a pay-streak.*

especially true in the richest portion of the pay-streak, but also towards the outside boundaries, and sometimes even outside of the pay-streak and below it. There can be a lot of black sand in a fine gold pay-streak. Sometimes the increased amounts of black sand extend out beyond the heavy concentration of fine gold and can be detected as itself in sampling (See figure 1-20).

Sometimes the color, hardness, and/or consistency of the streambed material will change towards the outside boundary of a pay-streak. This is not a true test of a pay-streak's outside boundaries though; because sometimes the streambed material will change inside of a pay-streak. Sometimes there will

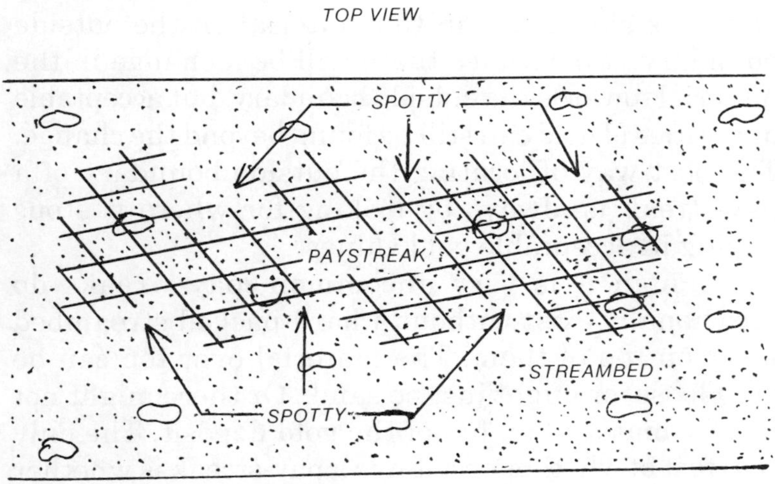


*Figure 1-19: Some pay-streaks are spotty towards the outside edges.*

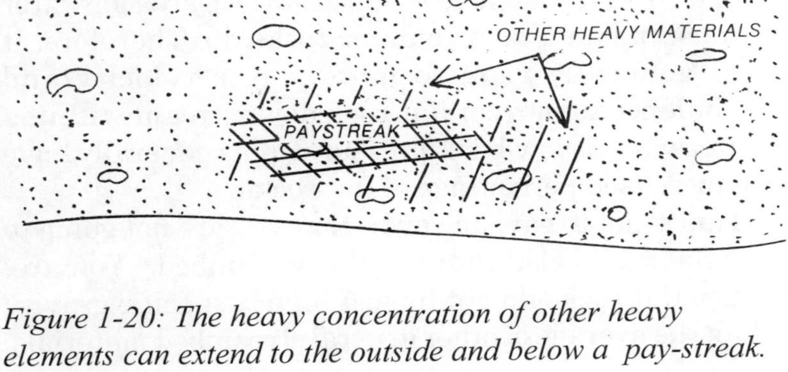


*Figure 1-20: The heavy concentration of other heavy elements can extend to the outside and below a pay-streak.*

be no visible change in the material on the outside boundary. Sometimes there will be a change in the material towards the outside boundary, but acceptable quantities of gold can still be found beyond the change. The best way of locating the outside boundary of a pay-streak is always by looking for where it stops paying in acceptable gold values.

Although it is very uncommon, pay-streaks do occasionally exist without a hard-packed streambed layer on top of them. The material over top can be very loose, or might just be sand. Or there might not be any material on top of the gold deposit. The only real test of whether or not it is a pay-streak is whether or not the gold is there! ***However, there is almost always a layer of hard-pack that rests on top of a pay-streak.***

So in sampling, it is wise to direct your efforts to reaching the bottom of hard-packed stream layers that are located where you believe the general gold path is traveling down the waterway.

## FLAT ROCKS

Flat rocks do not have as much surface area for water to push against as a rounder rocks do. Therefore, it will generally require more water velocity and turbulence to move a flat rock downstream within a waterway. So low/back-pressure areas generally have an increased percentage of flat rocks.

This is not to say, however, that you are not going to see flat rocks elsewhere in the streambed. You are. Since flat rocks do not have a higher specific gravity than the average of other general streambed materials, you are going to find them all throughout the waterway. Just seeing a flat rock by itself, or even an accumulation of them, is not a sure sign of anything. But if you already have an idea of where the general line of gold's travel runs in the waterway, and you are dredging sample holes along its length looking for the low/back-pressure areas, a visible increase in the percentage of flat rocks in the material can be a good sign to look for. Especially since you cannot always see a visible increase in black sand and iron rocks in the material even if they are there.

Sometimes during sampling, you can find all the correct signs; an increase of iron particles, lots of lead and old rusty nails and an accumulation of flat rocks - but no gold whatsoever, or very little of it. In this case, you have located a low/back-pressure area that is either off the main line of gold's travel, or is on the main line in a place where little gold traveled through during the major storm which formed that streambed.

A lot of lead by itself in the material or along the bedrock in a location might not signify a low-pressure deposit at all, but a good fishing hole or target-practice area instead. Look for the other signs of a pay-streak as well.

## NATURAL & VIRGIN STREAMBEDS

In my book, ***The Basics of Successful Gold Dredging***, I have thoroughly covered what a naturally-formed hard-packed streambed looks like, so I will not cover it again here. But it is important that every dredger understands what hard-pack is, because almost every valuable pay-streak you will find in a waterway will lie along the bottom edge of a layer of naturally-formed streambed.

The question is often asked, what is the difference between natural and virgin (I prefer to call it *"original"*) streambed? Original streambed is a naturally-formed hard-packed layer that has never been disturbed by man. It is a place that has never been mined before. You will not find any man-made objects underneath or inside of original streambed, although you often find them lying on top. All of the gold that has been deposited underneath original streambed still remains there today. We find a lot of original streambed at the bottom of some of the deeper rivers in California where the early-timers were not able or willing to go.

Natural streambed is any hard-packed layer that has been formed by a major flood storm. While this includes original material, it also includes any layer of streambed that has been formed since the rivers and streams were mined by earlier generations of gold miners.

It is really important to understand that a major flood storm will deposit new streambed and new placer deposits into any area which has already been gone through by earlier mining activities. Any area which formed rich pay-streaks in the past is likely to do so again if more gold is washed down into that area by another flood storm. The riffles in a sluice box will recover gold again and again after they are cleaned out, once you start washing new material and gold over the riffles. A proven gold-bearing waterway will react in the same way. ***So it usually makes little difference whether an area has been mined once***

***or twice already, as long as the mining has not happened since the last major flood storm which occurred in the area.***

Those areas which were once heavily mined by the early-timers were mined because they contained rich pay-streaks. Most often, the conditions which formed the original pay-streaks have not changed. Those very same factors which caused gold to concentrate there once, have likely caused gold to concentrate there again during any of the major storms which have occurred in the area since the earlier mining activity took place.

So it can be greatly to your advantage to know where the old-timers hit it good. The same areas are often paying gold dredgers just as well or better today. You do not have to find original streambed material to find a rich gold deposit. Most often, though, you do need to find hard-packed streambed. The dynamics that created the hard-pack will also have created pay-streaks underneath the hard-pack.

This is important: ***If you are finding hard-pack, then no miner has been there before you since the major flood storm which created the hard-pack and any gold deposits which may lie underneath.***

# 2

## SAMPLING IS THE KEY TO SUCCESS

***A hard-working, persistent, sampling activity is the key to long term, continuous success in any gold mining activity of any size.***

For over twenty years, I have personally devoted a lot of time circulating around within the general areas where I have been mining, gathering information about the waterways, talking to other miners, finding out where they have dredged and how much gold they were finding. I have had an opportunity to observe many dredging operations of different sizes and kind, some which have succeeded, many which have not. Out of all the operations I have seen fail, almost all of them failed because they did not sample effectively.

**Sampling is the key to success in gold dredging.**

The philosophy behind my own dredging success is actually very simple: When I am into a pay-streak, I work very hard to recover the gold out of it. When I am not into a pay-streak, I work very hard at sampling to find another one.

## DREDGE SIZE

Natural hard-packed streambeds are mostly made up of cobbles, rocks and boulders, with some gravel, silt and sand that fills the spaces in between. Progress in a sample hole is mostly made by moving the oversized rocks out of the way, those rocks that are too large to go up the suction nozzle. The primary reason why a larger dredge can get more done, is because larger-sized rocks can be sucked up the dredge hose. That means less rocks need to be moved out of the sample hole by hand. Please understand that I am not saying that you do less work using a larger dredge. I am saying that the same amount of work will get more accomplished with a larger dredge in experienced hands. Being able to suck larger rocks means the sample holes can be made faster, larger and deeper. This is important.

On the other hand, a larger dredge is generally more cumbersome to move around. It requires deeper water to operate efficiently. The components are bigger, heavier and require more work to move around and position against the constantly-moving water of a river. The motors are more expensive to operate. So using larger-sized dredge is a more substantial investment into a sampling program.

The idea behind sampling is to take a limited look at a prospect to discover if the area deserves a more substantial investment of your time and resources. There is a balance which must always rest upon your own judgment of how much to invest into a prospect without overdoing it. This needs to be compared against the other prospects which are available to you. There is not always an easy answer to this.

But often the answer about dredge-size is pretty easy: If I have a 5-inch dredge and an 8-inch dredge, and the prospect I want to have a look at requires the dredge components and fuel to be packed a mile over rough terrain, and I am not already certain there is a rich deposit there, or I only have a limited amount of time for the project, I am going to pack the 5-inch dredge in there – or perhaps even find a smaller dredge. You guys would do the same. It is a matter of making the most out of limited resources to answer the question about how much gold exists within a specific area.

In another scenario, if I have the same two dredges and want to sample a place along a larger-sized waterway where I can float the 8-inch dredge in, I will opt for the larger dredge.

Larger dredges will get more done in experienced hands. But they are more difficult to move around and set up. So you will have to decide about the preferred-size of your own dredge, based upon your own abilities, the remoteness of the area where you will operate, the estimated depth of streambed material within the waterway and how much time you have.

Once you decide upon your dredge(s), you will want to try and target areas to sample where you can be effective with that equipment. By this, I mean places where the streambed depth will not overwhelm your capability, and that are not so remote that getting the gear in there is going to turn into a bigger project than the prospect deserves.

A 5-inch dredge is an excellent all around dredging machine. It can be moved around easily, and is also able to produce well in a wide range of conditions when a pay-streak is located. A 4-inch dredge is also a good sampling dredge if you feel a 5-incher is too large to manage alone.

By the way, the size of your dredge does not determine how much gold you will recover. Of all the variables, I would say that the effectiveness of your sampling operation is the main factor that determines how much gold you will find.

To give an example of this, there is a guy I know who owns an excellent claim on the main river in the general area where I operate. His claim runs with a section of fast water in the center and slower water towards the upper and lower sections. At the beginning of a recent season, without having done any sampling beforehand, this miner floated his 10-inch dredge onto the upper section of his claim where he figured the gold ought to have been. He then spent most of the season dredging in that same particular spot, working his hole down to bedrock and continuing to spread it out, and out, and out - looking for the gold he thought should have been there.

Usually when you see someone dredging in the same place for days, weeks or months, you figure that the person must be doing pretty well in that spot. Not so with this guy. He was simply dredging continuously in that same place in hopes to eventually get into some kind of nice deposit.

The worst of it was that he also had a 5-inch dredge floating in the water on the upper section of his claim. The 5-incher was not being used at all. Each time that I saw him, I advised that he temporarily shut down the commercial dredge and spend a few days or a week sampling around the claim with his 5-incher to see if he might find anything interesting to dredge elsewhere on his claim - especially in and around the faster-water areas where his 10-inch dredge would not go as easily. Each time we talked, he agreed that my plan was *"probably a good idea,"* but he did not want to work the smaller dredge when he had the bigger machine available to him.

Towards the very end of the season, he finally realized that his 10-inch operation had failed, and took some of my advice by sampling the upper-end of his claim with the 5-incher. He did not find a pay-streak in that section of river. But that is something that he could and should have found out about at the beginning of the season.

He never did sample the lower section of his claim below the fast water. His reason was that it would have been too much of a hassle to move the 10-inch dredge into that section of river from where it already was located. That was true only as long as a rich pay-streak was not yet located there. It is never too much trouble to move gear around once a rich gold deposit has been found!

A short time later, the guy ran out of operating money and sold his 10-inch dredge to a more successful miner for about a quarter of its value. This was so he could have money to get home. He also agreed to allow another dredger go onto his claim on an exclusive ten percent lease (10% to the claim owner). This new miner floated his 6-inch dredge upstream from an access-point located further downriver, to a place well below the fast water (on the claim where the guy with the 10-inch dredge failed). His plan was to start at the lower-end of the claim and work forward with a systematic set of sample holes in an effort to find the pay-streak if it was there. And BINGO; he hit it on the first sample hole!! It was a rich pay-streak that paid two-to-three ounces per day to a 6-inch dredge in about 2 1/2-feet of hard-pack over bedrock. The 10-inch dredge would have done really well there!

***Sampling is the key to successful gold dredging.***

I can never say it too many times!

Every situation is different. The best answer is to choose places where you can sample effectively with the gear at your disposal. It is unwise to commit all your resources to a single prospect until the gold deposit has already been confirmed by sampling.

## CHOOSING AN AREA TO SAMPLE

***Placing your dredge operation into an area that contains rich gold deposits in the active waterway is the first and most important step.***

No matter how good your gear is, or how hard you work at an effective sampling program, if there is no gold there in the first place, you are <u>not</u> going to find any!

If you do not already have access to a good section of river, or do not already have some hot leads where to go, one of best things to do is to locate an area which has already proven to be <u>very</u> rich in the past.

Past history of rich lode mines in the area does <u>not</u> always mean there are rich placer deposits within the waterway. Past history of rich river or stream deposits usually <u>does</u> mean they exist there again today. There are several methods of doing this. One way is to get hold of a ***Where to Find Gold*** book to identify the different locations in the areas of your interest which have proven to produce rich stream-deposits in the past. These books are available from the numerous prospecting equipment suppliers around the country. Geological reports and a whole variety of maps are available to help you evaluate different general areas and specific locations in your search for a good place to start. These reports are available from the Counties of your interest, or the State Geology & Mining Departments. A lot of this data is now available on the Internet.

Another way to find a good gold-bearing location is by joining a mining organization, and/or the various mining associations near where you live or in the areas where you want to mine. There are many such organizations. Some of them are in the business of aggressively obtaining extensive gold-bearing property for their members to work.

Some mining associations have many miles of mining property available to their members, and add more as time goes forward. Each group has its own set of rules. Different clubs and associations have mining properties available to their members in different areas. For example, ***The New 49'ers*** in Northern California makes 75 miles of mining properties freely available to members along the Klamath, Salmon and Scott Rivers in northern California. For about the cost of buying a single mining claim, members gain access to more gold-bearing property than they can explore in a lifetime, can mine and camp as much or as often as they like and can keep all the gold they find. Most mining clubs and associations today can be found on the Internet. The New 49'ers are located at www.goldgold.com.

Once you do choose a general area where you will operate and go there, you should spend some time looking the river over to evaluate which sections are most likely to contain rich pay-streaks. One of the best ways to do this is through communication with the locals and other miners in that area, to find out where gold has already been taken out by dredging activity.

Some sections of river are far richer than others. A rich section of river is present wherever rich gold deposits have already been taken out with dredges here and there along its length. If paying deposits have already been taken out of a stretch of river by suction dredgers, it is nearly certain that there are more undetected pay-streaks remaining in place.

It is usually no secret where the paying deposits have already been taken out. In fact, this kind of information is usually broadly known, even if the specifics are exaggerated or understated. Your best bet is to get your data from reliable sources - like those who recovered the gold, or if not available, by as many different sources as possible. Then you can at least get a general idea of what happened and where. It does not really matter if the accounts (how much gold was recovered) are 100% accurate. What you are basically trying to do is locate those sections of river where you know that paying quantities of gold traveled through and deposited during the last major flood storm. Once you begin finding out about the specific areas where gold has recently been taken out, you can start planning where to focus your sampling activities.

***The idea is to go for the sure thing:*** You pick an excellent past-producing general operating area. Then you locate specific areas along the waterways in that area where gold dredgers have been finding pay-streaks. This allows you to go into those sections and actually see where the general paths of gold are most likely to run. Then you can make a plan to sample the obvious and likely low/back-pressure areas along that path. This procedure will put you into high-grade gold deposits quickly.

Sometimes you can pick up some information about a gold deposit being located in a specific place – the deposit which is still present. This happens on a regular basis in mining clubs and associations when different members only have a limited amount of time to work out deposits once they are found. Once they leave, the deposits get picked up by other members. We see this all the time in ***The New 49'ers.***

When you are making a sample plan, based upon the information you have been able to collect about the area, you are best off to go after the most likely locations where the deposits ought to be. This is similar to aiming for the bulls-eye on a shooting target.

There are other ways to go about finding gold: You can just put your dredge into the river anywhere and start dredging in hopes that you will discover a gold deposit someday. I have never seen this method work very well. It is a big world out there! It is better to target the areas where gold is more likely to be, rather than just choose anywhere and hope for the best.

Or, you can go out and sample different streams where nobody has ever found any gold, hoping that perhaps those areas were overlooked by earlier prospectors. It will not take very much sampling in such areas to gain an idea of the prospects. Most streams where there is little or no gold present will not even produce a single speck of gold for your efforts.

You could check out those areas where streams have shown a small amount of gold during the past, believing that you might be more likely to find a concentration, since fewer miners have worked those streams. It is wise to pay attention to your sample results. They answer these questions.

There are probably many different ways to go about finding paying quantities of gold, all with a certain degree of workability. However, if you apply yourself to the method which works most often, you will be increasing your chances of continuous success by that much. In my experience, the most consistently-successful way of finding and recovering paying deposits of gold is to go after the sure thing, or as close to it as possible. This means going to areas that are already proven by past mining history, and then finding the sections of waterway where modern-day miners are locating pay-streaks that have been formed during the more recent major flood storms.

Once I determine where the paying deposits are most likely to be. I will make a list with the best prospects at the top. All the items on the list will be good prospects. Then I will contact the claim owner of the first prospect (if there is one) and see what kind of a deal I can work out. If an acceptable deal cannot be worked out on that section of river, I will try to make a deal on the next-best prospect on my list, and so on, until I have a place to mine legally. You can usually make a deal with most owners of mining claims. If they are not working the claim themselves, they will usually be interested in allowing someone else to work the claim, giving them a percentage of what is recovered. If you cannot come to an acceptable agreement with a claim owner, there are plenty of other places to dredge!

The best opportunities available to most prospectors today are with local and national gold mining associations/clubs; especially those that are aggressively acquiring mining property for their members to work in the proven areas. With lots of property freely available to you through a mining organization, you do not have to negotiate with claim owners. You can just make your list of hot prospects and go to work.

# GATHERING DATA

There are some gold dredgers out there who have a fixed idea that they must mine only on their own claims, even if there are little or no paying quantities of gold to be found there. This is too bad really; because it can be far better to pay out 10% of a good find to a claim owner, or join a mining organization, rather than work a low-grade section of waterway on your own mining claim.

Every gold deposit eventually plays out, and that's when it is time to go find another one. This is why you will want to be continuously collecting new information about the surrounding area, even when you are dredging up a rich pay-streak.

There are many ways to collect new information about the general area that you are operating within. Paying attention to where others are dredging sample holes is one of the best ways. It is a sign of experienced dredgers if you see them dredging sample holes up through a section of river, and the holes appear to be mostly down to bedrock and they appear to be in the right places. So you can figure that they are doing a pretty good job of sampling that section of river and that they probably have not found acceptable quantities of gold, because they are still moving around.

In most rivers, the water will be clear enough that you can see all or most of the sample holes which have been dredged during the present season. Winter storms will normally fill in all the holes. So you usually can only see the open excavations of holes dredged during the present season.

If you notice that an operation is continuing to dredge sample holes, it is reasonable to assume they have not yet located an acceptable deposit. Or, if a dredge is producing regularly in the same spot, and you can see signs of their earlier sampling activities in the vicinity, you can figure that they are now into some kind of deposit. If the deposit is not producing for them very well (although it might still be a rich deposit and they don't know it), you probably will not see them produce there for very long.

***The proven existence of any gold deposit inside of a waterway gives you valuable insight about where the general gold path is located.***

When you talk to dredgers, if you cannot find out from them how well they are doing, you can at least find out how experienced they are just by asking questions and feeling them out. Then you can assess their experience (meaning their ability to find and develop gold deposits) against what they are doing, and get a better idea of whether or not they are into a high-grade deposit or not.

It does not matter nearly as much to know how well the other operations on the river are doing, as it does to know where on the river that other dredgers are finding gold deposits. This is the most important information you can collect: knowing exactly where the pay-streaks are being, and have been found. When you see where others have been finding gold deposits, you can immediately see where the general line of gold's travel moves through those same sections of river. And if you have an eye for the low/back-pressure areas, you can also see where other pay-streaks are most likely to remain in place.* Any section of river that carries one pay-streak, is more than likely to carry more pay-streaks in that same section.

The figures on how much gold another dredging operation is recovering out of a pay-streak can actually be misleading if you are not careful. The reason for this is that, for an accurate assessment of true value, you must relate the amount of gold to the amount of effective volume that is being processed by the operation. You cannot accurately assess effective volume without actually spending time underwater with the other dredgers. Less-experienced operators often waste a lot of time nitpicking around in a dredge hole, sometimes getting themselves all bound up with larger rocks and boulders, and sometimes dredging much deeper than they should to recover the gold off a contact zone up in the streambed. There are a lot of things dredgers can do to drastically lower the amount of gold that is recovered at the end of a production day. So you should rely less upon production statistics (unless they are good ones), than you should on identifying the proven places along the waterway which establish the path where high-grade gold has traveled and deposited.

For example, not long ago, I was listening to a member of our local mining club complain that he was only recovering 2 pennyweights of gold (1/10th of an ounce) per day using his 3-inch dredge along a specific

***Most of the common low/back-pressure areas are fully outlined and illustrated in GOLD MINING IN THE 21st CENTURY.***

area of the river. From the way he was talking, I believed it was likely that the guy was processing very little volume with his dredge, and probably dredging deeper than he needed to go to recover the gold out of a pay-layer. Shortly after he left, I encouraged one of the club's more experienced dredgers to go in with his 5-inch dredge and have a closer look. Sure enough, some more efficient volume-production only down to the pay-layer started producing 3+ ounces of gold per day in the very same dredge excavation. Other club members soon caught onto the discovery; and within a matter of weeks, there were a dozen dredges working out the rich discovery.

So you should not rely too much upon the production statistics of other operations, unless you can go down and see for yourself what they relate back to in terms of effective volume-production. It is enough to know where others have located the high-grade path of gold. It is even better if you can also establish the layer in the streambed where the high-grade gold is traveling. You cannot just assume the gold is on the bedrock, because there are many times when it is not. Mapping the gold path in this way will allow you a good perception of where your own sample holes will be most productive.

Another excellent source of information is the local gold buyer. He will know who is finding gold, because he will be buying it from them. Some gold buyers keep quiet about who and how much. Some are open with their information. It mostly depends upon the relationships that are involved.

Another way to come about good information, which will sometimes get you onto a good claim fast, is to offer a *"finders fee."* If you make it known around the general area that you will give 5% of your gross recovery to anyone who arranges to put you into an acceptable gold deposit, you never know who might emerge out of the local woodwork with some helpful information. After you pay it, and then offer the finder's fee again, you might get into another deposit just as fast.

When I do sampling projects in the remote locations of other countries, I have found that it sometimes works very well to offer a fixed financial reward for every hot spot that I am directed to by local villagers. There are usually locals in any gold-bearing area who are aware of where the high-grade gold has been recovered out of the waterway. Sometimes, the only chance of their gaining in any way from their knowledge is to sell it to someone else who has the capability to develop the mining potential.

Another good way to collect valuable information is to talk to every active miner in the area who will give you the time of day. This is not to say you should spend all your time running around *"shooting the breeze"* over a beer or a cup of coffee. Although this might be justified when you first come into a new area and you are still trying to establish where to start sampling.

Once you get started in sampling and production-dredging, you are naturally going to run into other miners in the area. If you can swing it, talk to them and find out what they know. If someone knows specifics, like: *"Joe Blow hit a bonanza down on Big Bend last summer,"* and *"John Doe found some great deposits down at Hurricane Flat,"* and if the person is willing to take the time, have him show you exactly where along the waterway that these gold discoveries took place. It is better to not assume you know where the person is talking about. The exact amounts taken out will probably not be wholly accurate, but that does not really matter. You mainly want to find out as best you can where all the proven sections of river are, where the general path of gold's travel lies in those sections, and what layers in the streambed are producing the high-grade gold.

The best sources of information when doing this are the miners who were directly involved with the individual gold finds. But if there is somebody there right now who is offering you information, take what you can get.

Here is an example of common questions that I might ask when gathering data from the other miners in my own general area:

*Where has lots of gold been found on the river in past history?*

*Where have gold dredging finds been made? Where exactly? How much gold?*

*Is the gold coming off the bedrock? Or from the streambed material?*

*How many layers make up the streambed?*

*How deep is the material?*

*What color is the pay-dirt?*

*Is it fine, medium or coarse gold?*

*What areas on the river have been heavily dredged by others? How much gold?*

*Percentage deals on the claims? How much? Who to talk to and where to find them?*

*What's the most consistently-paying section of river today?*
*Other good waterways in the area?*
*Places to camp?*
*Any problems with locals or government officials?*
*Any theft or crime problem in the area?*

When collecting information about the area, it can be a big help to have along an enlarged section of map which covers the river. Mark in the good areas as you find out about them. This is especially helpful when you are new in an area and are not yet familiar with the river or the names of places when locals are talking to you.

It is good to try and keep tabs on as many miners, claims and areas as possible. In this way, you will be continuously accumulating new information and updating your assessment of different prospects. This will allow you to always have a multitude of *"next best"* sampling locations on your list of likely places to find paying quantities of gold.

One thing worth mentioning at this point is that I personally never withhold information about the river from other miners who ask me questions. Let me clarify that: If I know where there is a rich deposit that I intend to dredge out at some later time, I might not say very much about it. And I never betray the trust of others when they have confided in me about their own deposits. But that is about as far as my own quietness goes. I figure that anyone's success in a dredging operation depends upon his or her ability to sample. The ability to sample depends largely upon how much information that he or she can collect about the operating area. I go about data-collection with abandon. In exchange, I give data out freely. And in doing so, I have never had any problems in finding out what I want to know. What goes around, comes around!

## "NO GOLD" REPORTS

You should not rely very much upon the information when you are being told that there is no gold left in an area. Although, you want to pay close attention to where sampling and production dredging has already been done by others.

Pay-streaks are not visible from the surface, although many of the places where they ought to be are visible to some experienced miners. Since the pay-streaks are not visible, and since nobody has found them yet, it is very easy and normal to assume that they are not present – even when they are. And so it is not uncommon to receive inaccurate reports from the local miners that there is no gold remaining in certain locations or sections of river. Just because others believe there is no gold, does not make it true! If you are smart about it, you will pay close attention to exactly where the gold has already come out of the river, and then make up your own mind whether or not certain sections of river ought to have paying deposits remaining in place.

Just by the nature of some pay-streaks, someone might dredge a sample hole down a few feet off to the side of a good one, see no signs of it, and miss it entirely. Quite a few times, I have seen guys dredge right down into the center of rich pay-streaks and miss them entirely. *"Not very much gold over there,"* when there is a hundred ounces ready to be recovered! An experienced dredger will start asking this dredger some questions about that area, the streambed material and the nature of the little bit of gold that was found, and then go right on in there after the sure thing!

I have also missed my own fair share of pay-streaks because of poor sampling procedure. As just one example, I was sampling a section of river a couple years ago and floated my dredge down into a large low/back-pressure area behind a large bend in the river. I knew there was gold there, because someone else had been into the location before me and told me about it. Plus, all the signs pointed to a very hot pay-streak sitting there. More than a thousand ounces of gold had been dredged out of the claim just above. The adjoining claim down below was producing just about as well. I had already recovered a 20-ounce pay-streak out of a much-smaller low-pressure area just downstream. I also knew from talking to the guy who was in there before me that the material was only two or three feet deep, that he did not do very much work in there and that his best day was seven ounces with his five inch dredge! Now that I look back on the whole thing, the signs appear so obvious that there might just as well have been a blinking neon sign over the area with an arrow pointing and saying: *"Rich pay-streak here!"* But at the time, the signs were not that clear to me. I just had an idea.

So I floated my dredge down into that spot. The water was swift; and because I was working alone, it was very difficult for me to get my dredge set up into position to dredge a sample hole there. I worked hard at it for about an hour, struggling against the current, making progress, losing it, gaining it back again, and so forth; and I finally did get set up in that location. The earlier miner's hole was still there. All I really had to do was continue it for a bit just to take a look. But I was exhausted and disgusted from the exertion of setting up, and I didn't want to leave my dredge in that position all night. There were a few hot leads I wanted to sample further downriver, and so I never did dredge that sample hole. I floated my dredge downstream, instead.

About a month later, some other dredger who was operating off of the information he had received from me moved into that location and dredged that same sample hole; the one that I didn't do. He ended up pulling out better than 120 ounces of rice-sized gold and nuggets inside of three weeks! My sample hole would have been right smack in the middle of the paystreak had I gone to the extra trouble of dredging that hole, which would only have taken me about 15 minutes. That gold would have been mine!

This story should make you wonder how many paystreaks are floated right over by experienced gold dredgers, not to mention the ones that are missed by oversight and poor sampling procedure.

When someone does not find high-grade gold in an area, everyone just naturally assumes it is not there – even when it is! So when someone says that there is no gold in an area, it just means that there is no gold in the area that the person knows about. This is probably the best way to look at it.

There is one other thing to also keep in mind about this: There are only a small percentage of gold dredgers out there during the time of this writing who understand how to sample effectively, and who actually do so. Consequently, there are unsuccessful gold dredgers around the various proven gold-bearing areas. There is no better excuse for an unsuccessful gold dredger to use than: *"There is no gold left in the river'."* Some failed gold dredgers will go out of their way to convince you that there is little or no gold left in the river. Because if you, being new to the area, start finding paying deposits there, what kind of a miner does it make the failed gold dredger? You are better off to not judge by the negative reports of failed or inexperienced miners.

You can get far more accurate data from the continuously-successful miners, but even the negative reports from them are suspect. Maybe they don't want others moving into their general area of successful operation? People are just not continuously successful when there is *"no gold."* Find out what has already been done, and found, and where, and use your own best judgment about what it all means.

## EXAGGERATED DATA

You also have to watch out for exaggerated information. There have been <u>many</u> times when I have gone to town and heard reports of my own recent gold finds which had been blown up to three, four or even as much as ten times larger than the finds actually were! This happens; it is all part of the game, so you have to expect it. As mentioned earlier, the exact amounts do not matter nearly as much as knowing a discovery was important, and pinpointing the exact location of the find.

The trouble with exaggerated data, is when you hear of a *"tremendous," "incredible," "amazing"* find from another general area some distance away, it tends to make you want to *"rush"* on over there to start dredging. Before deciding to pick up and abandon the general area you are operating within for a new general area, it is a good idea to try and substantiate the reports to the best of your ability. Who knows; the reports that you are receiving about that distant area might be small compared to the reports that they are receiving about your finds in your own general area of operation!

There is nothing wrong with leaving one general area for another if it is a better one. However, it usually takes being in an area for a while before you gather enough understanding of it to where you can evaluate accurately where the deposits are likely to be and start putting yourself into them rather continuously. This is a lot to give up because of an unconfirmed report or two!

On the other hand, there is the chance that after prospecting an area for a while, your greater understanding of it may show that there are not many deposits to be found and you will need to find a new operating area. So don't shut out those rave reports and success stories either; they might be just what you need!

# SAMPLING PROCEDURE

How many sample holes that it will take to get into a good pay-streak depends largely upon how good the general operating area is, how well you place your sample holes and your thoroughness in following up on positive signs while sampling.

It might take a few more holes to find the first pay-streak in a certain section of river, than it will to find the additional pay-streaks in that immediate area. Once you find one deposit, after it is cleaned up, your sampling is likely to turn up additional pay-streaks in the likely spots along the same line of gold's travel in the same immediate area.

When dredging a sample hole in a likely spot, it is usually a good idea to dredge one or two more holes off to each side, to make sure you do not miss the pay-streak if it is present. This includes taking a sample of the waterway closer to the bank if it is right there, and also dredging a hole or two out towards the gut (center of river), especially out into the gut, as long as you can accomplish it safely.

You should keep your sluice box relatively clean of gold while dredging sample holes. Sometimes you might pick up a few pieces of gold out of a sample hole that you might not see while dredging. If you already have some other gold accumulated in your sluice box, you might not notice the difference. Sometimes just a few additional pieces of gold in your box may be the only sign that you are sitting right next to a rich pay-streak. So it is very important to be able to tell when you have any amount of additional gold in your box. Sometimes you will accumulate some gold in your box from earlier sampling in areas that had some gold, but not in paying amounts. If the accumulation becomes enough that it will interfere with your ability to see additional gold in the box even in small amounts, you are better off to dump the concentrates into a wash tub and then dredge them up again later when you are ready to do your next clean-up.

Most suction dredges today have a special high-grade section of riffles at the head of the box where most of the gold is recovered during dredging. It might only be the first 5 or 6 riffles in the box. This special high-grade section of riffles can be removed and cleaned up quickly after every sample hole. This saves all the time and effort required to clean up the entire recovery system every time a sample is completed.

A few pieces of gold seen on the bedrock, in the streambed material, or ending up in your sluice box after dredging a sample hole, is one of the best signs to look for of a nearby pay-streak. This is especially true if the pieces of gold are larger than fines.

If you are seeing an increase of gold from a sample hole than what would normally be expected in the general material outside of a pay-streak, your sample hole should be moved to both sides and forward, with you looking very closely to see if you can find more of the gold, and see in which direction it seems to get better. A few pieces of gold seen while dredging a sample hole are always worth checking out further.

If you do not see any gold while dredging a sample hole, but find some new gold in your sluice box after the dive, it is always worthwhile to go back down and open up the sample hole further while looking closer, until you find where the gold is coming from and how much is there.

***The smallest clues often lead to the richest pay-streaks!***

If in dredging further, the amount of gold does not seem to get any better, try dredging a few more sample holes in the immediate area to see what else you can find. It may be that you are near a rich pay-streak. Or it may be that you are already dredging inside a low-grade pay-streak. Or it may be that you are along the general line of travel, and not yet into a pay-streak.

Sometimes, with further dredging in the same sample hole, you will see more gold along the bedrock or in the contact zone between flood layers, but you still cannot be sure if the area is rich enough to dredge on a production-scale. In this case, it can be worthwhile to spend a dive or two, or the rest of the day, production-dredging in that spot, and then clean up your recovery system to see how good it is going to pay. This is best done after additional sampling in that immediate area has already shown that you are not going to find anything much better around there. Sometimes this happens when dealing with widely-dispersed fine gold which is spread throughout the material. The gold can be so fine and spread out, that you are not able to see much of it while dredging. Sometimes fine gold pay-streaks such as this can pay quite well on a production-scale.

Some rivers are so heavily saturated with fine gold that you have to clean out your sluice box quite regularly while sampling to observe the increase in gold after each sample hole.

Once you have finished cleaning up a pay-streak, after your first sample hole, sometimes you may find some gold in your box that you are reasonably sure did not come out of the new sample hole. After working in a good pay-streak, some gold can become trapped in the classifier and other places inside your suction hose, power-jet or header box. It does not usually take very long for this gold to work its way free and then show up in your riffles. This usually happens on the first sample hole or two. However, it can be misleading to you while you are sampling if you are not aware of it.

When choosing places to dredge your sample holes, pay attention to where the large-scale earlier mining operations took place. Many of those old-time operations, particularly the hydraulic mining operations, washed a lot of gold into the river. Watch for where their tailraces entered the river. Do some sampling right there if you can. When these do pay, they usually pay very well. Sometimes the gold from these tailings has been washed downriver in a major flood storm to form an additional common line of gold, separate from the main one which runs down the river from further upstream. The same holds true for high bench bleed-offs.

If you watch for these places and dredge a few sample holes downstream of them where you judge that the line of gold might run, you can find yourself dredging up pay-streaks and single deposits along the *"wrong side of the river,"* sometimes even on the outside bends!

Usually though, when you follow a line of gold to the outside of a bend, you will find single deposits of the larger-sized pieces of gold where the changes in bedrock and other obstructions allow for them to get trapped. Outside bends, for the most part, are not good places to locate pay-streaks, because there is such an increase of force there during a major storm. This is not always the case though. If it is a mild bend, and circumstances allow for a low/back-pressure area there, and there is a source of gold emerging into the river not too far upstream, you just might find a well-developed pay-streak on the outside. I have seen this happen numerous times.

So it is not just the deposits on the main common path that you are sampling for. If one of these other conditions presents itself, it is worth an extra sample hole or two to check it out. *"Nothing ventured, nothing gained,"* really does apply to sampling.

## READING STREAMBED MATERIAL

As mentioned earlier, sometimes the gold will end abruptly at the outside of a pay-streak. Just as often, though, the amount of gold keeps getting less and less further out from the richest portion of a pay-streak. So there are usually some signs that a pay-streak is present nearby if you happen to dredge your sample hole to the outside and not too far away from the pay-streak. If you see these signs, they can usually be followed right into the pay-streak.

It is <u>important</u> to keep your eyes open for these signs when you are dredging sample holes. Looking for these signs is your <u>entire</u> purpose of dredging the sample holes in the first place! Not only are you looking for gold, but you are also watching for those places where you are most likely to find an accumulation of gold if it is present. The contact- zone between two layers is a very common place to find gold if it is going to be present. The same is true for bedrock. You will want to be looking for those places along the bedrock where gold would definitely become trapped if a pay-streak is present in that location – like deeper cracks and holes which create excellent gold traps. Uncover these particular areas slowly and look closely for any accumulations. These are the most likely places to find gold if it is going to be there.

If the gold traps exist there under hard-packed material, and you cannot find any gold inside of them, you know you are probably off the established gold path. So move your sample hole or excavate a new one off to one side or the other.

You are also looking for accumulations of rusty objects and lead in these very same places. They are a sign that you are dredging your sample holes in the right locations, or near to them. Increases in natural magnetite (iron rocks) can be a good sign too, although you might not notice them until you finish the dive and look into your sluice box.

In some rivers, some colored materials pay much more often than other colors, like a red color. This is not always true. But if it is, it just gives you one more sign to zero-in on.

If you start seeing an accumulation of the heavier minerals, rusty objects or lead along the bedrock, or in the contact-zone between flood layers, you should start spreading your hole and watch closely to see in which direction there seems to be more of it, and at what level in the streambed material it is located.

Watch closely to see if there is any gold present, too. Seeing gold is the best sign. Once you start seeing gold, you look closer to see exactly where it is coming from out of the material. Then you start looking to see in which direction the gold and other signs seem to get more plentiful. Move your sample hole in that direction, always uncovering the area (where it seems to be paying the most) slowly enough that you can see more if it is there. And pretty soon, if it is there, you will be into the pay-streak.

There are several examples I like to use when teaching people how to sample for gold. The first is like hound dogs on the trail of some person or animal. In this case, let's say the dogs are trying to track down a person named Jack who is attempting to get away. We all know that every time Jack touches the ground, he leaves a scent behind which the dogs can detect. I don't know exactly what substance is that is being left behind; but for the purpose of this discussion, it does not matter. It is enough to know that whatever it is, once the hound dogs are directed to follow it; as if their whole existence depends upon it, with single-minded determination, the dogs continue to sniff around for every step Jack took in trying to get away. As long as Jack's route is confined to the ground, he will leave signs behind which the dogs are usually able to follow.

The route of gold as it is washed down a waterway during a major flood storm is <u>very</u> similar to this. Because the gold is exerting a downward push that is about 6 times heavier than everything else, almost all of the gold is washed downstream across the surface of whatever structure remains solid at the bottom of the waterway. Sometimes this is bedrock. Other times it is an old compacted layer of streambed which is in place above the bedrock. Traces are left behind as the gold is washed across this fixed surface-structure. Generally, the traces are <u>much</u> more abundant on the general gold path, than off of the path – because so much more gold has traveled across that particular surface. ***This is a comparative statement.***

While dredgers cannot sniff out the gold trail as readily as a hound dog can follow the scent of a man who just ran across an open field, through trial and error, a dredger <u>can</u> locate and follow the path which most gold follows down the waterway. That path <u>does</u> have the most gold traces, even when they do not amount to a high-grade pay-streak. Your recovery system is able to trap most of this gold. So, through a series of test holes, and by paying close attention, you can see where most of the gold is traveling down along the waterway. With few exceptions, all of the high-grade gold deposits will be found along this very same trail in the river. Once you get yourself tuned into what you are looking for, the signs are not so difficult to follow.

Ultimately, you are trying to find the rich deposits where the gold is so abundant that it is richly-visible throughout the pay-streak – just like the dog when he finally catches up with Jack. But leading up to that, sometimes the first signs that you have to look for in the streambed are where the rusty objects are located, and where there is only a very small increase in gold over background.

What does that mean; *"over background?"* It means that a certain amount of gold will be present throughout any gold-bearing waterway. The amount is usually very small. You will get an idea what this is once you have dredged several sample holes in any given area. What you are looking for is some significant increase over the background. That is the scent you want to follow.

You must be careful what you compare to during the early phase of a sampling program. Some impatient miners compare what they are finding during sampling to what they would like to find in a rich pay-streak. That comparison is not helpful and only leads to frustration. The dog is comparing where Jack has stepped, to where Jack has not stepped, so he can follow Jack's trail. Likewise, in sampling, you must follow where there are more traces of gold in the waterway, compared to where the traces are almost nonexistent. Sometime the difference is very slight. So you have to really be patient and careful in your process.

The other example I like to use is that of a drug addict who is hooked on one of the most euphoric substances available to mankind (gold qualifies for this, although that's not what I am talking about here). Most people today have had some contact with someone they know who would do anything to get another charge from a narcotic drug; steal from their family and friends, turn their backs on those who depend upon them, perhaps even kill. Am I right?

Getting into a high-grade gold deposit triggers similar feelings of euphoria. Different people deal with it in different ways, but everyone gets the good feeling. It is just a matter of how much gold it takes to trigger the reaction. ***The reaction is called "gold fever."***

As you are mining up a pay-streak, you will find yourself slowing down to look every time you get to that place in the pay-streak where the gold is richly concentrated. This is because seeing all that rich, yellow gold being uncovered for the first time really gives you a big emotional charge. It feels great!

But it also feels terrible when you get down to the layer where the gold is supposed to be, and there is no-longer any gold there. This always happens at some place on the outer boundaries of the pay-streak or when the deposit eventually plays out. With a little experience under your belt, you will develop a hunger to re-experience your own euphoric feelings that are energized as you uncover high-grade. Through this, you will naturally find yourself following the richer part of the pay-streak, and avoiding the areas to either side where there is not enough gold to produce those wonderful feelings that you want to experience again and again. And so following the pay-streak becomes a natural affair.

Good prospectors are the ones who have developed a similar sensitivity to what I am talking about here, but on a much more magnified scale. They get a charge out of detecting the much-smaller differences in how much gold is present on the gold path outside of the high-grade deposits. Just as the hound dog must experience some feeling of satisfaction every time he finds another place where Jack touched the ground, a good prospector feels the progress in his sampling program as the low-grade portion of the gold path is followed up the river.

No-matter how well trained and disciplined any dog is, or no-matter what kind of punishment is threatened, we all know how hard it is to get a dog's head out of the bushes once the dog has picked up some interesting scent in there. Am I right? This is the very same single-minded determination that good prospectors use to follow the gold path in search for high-grade gold deposits. Just as Jack is guaranteed to be somewhere ahead along the scent he has left behind for the dog to follow, high-grade gold deposits will be found along the path which most gold travels down along the waterway.

Reading the streambed material in sampling is the process of following the lower-grade gold trail and other heavy materials which will eventually lead you into the high-grade gold deposits.

## DEPTH OF MATERIAL IN SAMPLING

I have run across quite a few dredgers who believe that the streambed material has to be deep to produce rich pay-streaks. This is simply not the case. Many of my own important discoveries have been made in three feet of hard-packed streambed material or less. I have seen or know of literally thousands of ounces of gold that have been recovered by suction dredgers out of three feet or less of streambed material.

I have also seen a lot of gold recovered down at the 10 to 15-foot levels ( of streambed material ).

All rivers seem to have their deep sections and their shallow sections. Every river that I have been on has had its deep and its shallow gold deposits. In fact, I recovered a pay-streak of black sand and fine gold in Alaska which ran right down across a section of exposed bedrock. There was no material over it at all!

The deeper that a pay-streak is located under streambed material, the richer it will have to be for you to recover the same amount of gold on a daily basis using the same production equipment. This is because it takes more time, expense and effort to move a larger amount of streambed material to expose the gold deposit.

It is wise to direct your sampling activities towards those areas which you can effectively dredge with the production equipment at your disposal. If your effective production capabilities are in six feet of material or less, why dredge sample holes greater than six or eight feet deep? One good reason would be if the sample hole is being put down in a prime location and all the signs of an excellent deposit are very positive. Once in awhile, you will find yourself in this predicament, and you will have to use your own best judgment when deciding how deep you are going to dredge a sample hole before pulling out of it.

Using a 5-inch dredge, I can usually dredge three or four good 4-foot sample holes in about the same amount of time that it would take me to dredge an effective 8-foot sample hole into hard-packed material. So what is better; an increased amount of effective sampling at shallower depths? Or, to sample fewer areas at greater depth? There is no fixed answer to this question. Every situation is different. It depends upon how good the prospect looks. You will have to decide for yourself.

Certainly, if you cannot work the deeper areas effectively, then you should concentrate on covering more of the shallower likely spots with your sampling plan. The problem is that you often cannot tell how deep the streambed material is going to go before you commit yourself to a sample hole. So what if your sample hole is going deep, but all the signs are very positive? This is something dredgers have to struggle with all the time. You will have to decide yourself, because each situation is different.

Having said all that, you have to be careful not to get pulled too far into a single sample hole. To finish a sample hole just because you started it, is not good enough reason to go well beyond your effective production capabilities.

I am quite certain that there must be gold-bearing rivers which run deep for the most part, and so most of the deposits will run deep, too. You will find this out once you start sampling and collecting data about the area. If this is the case, and you do not have the equipment to effectively handle the depth, you may be better off to choose a different general operating area where the rivers have more shallow areas to dredge. There are many of these to choose from.

Deep areas are easiest sampled with larger equipment. When figuring out the amount of time and energy it would take to dredge a 15-foot sample hole into streambed material using a 5-inch dredge, you probably could not go to too much trouble to float an 8-inch dredge into the location to complete the excavation.

Actually, in a sample plan, you should not be thinking in terms of a single excavation. You should be planning to cover an area with multiple excavations. This is all the more reason to not get in too deep beyond the effective depth-capability of your sampling gear.

The deeper you go beyond the effective depth capability of your dredging equipment, the richer the gold deposit must be to justify all the extra time and effort. But richer deposits exist in fewer numbers. This means the deeper you go beyond the effective limit of your gear, the less chances you have of a successful venture.

Please don't misunderstand this last statement. I'm not saying the deeper you go, the less chances for success. I am saying the deeper you go beyond the effective capability of your dredging equipment, the less chance you will find enough gold down there to make dredging with that gear worthwhile. Sometimes, when the prospect is encouraging, we dive as deep as we can go, with the idea of bringing on larger equipment if we find something good.

So when looking over and gathering data about a section of river, you should also be very concerned with depth of streambed material. Any good sampling plan must project the average depth of streambed that you will need to penetrate to complete the samples.

It is good to have some plan of how deep you will take a sample hole before pulling out of it. Different dredgers have their own methods of determining this, and you will have to decide upon your own. Just to give you something to go on, though, I will tell you what I do: I figure that I can dredge effectively down to about six feet of streambed material with my 6-inch dredge (five feet with a 5-inch dredge, eight feet with an 8-inch dredge, etc.). I can dredge deeper than this. But it has to pay very well to make it worth my while, because I find so many pay-streaks at less than half that depth. Since I am comfortable dredging in six feet or less, I will dredge a standard, safe, good-sized, orderly hole (this procedure is covered in ***Basic Gold***

***Dredging Techniques***) down to about the five foot level of material. If at that point, there is still no sign of gold, a change in flood layers or of bedrock, I will make a dive straight down for bedrock at the bottom of my hole. By this, I mean dredging a small hole - about two or three feet in diameter - straight down (called a *"coyote- hole"* by dredgers) just to see if I can find bedrock or a contact zone between layers. This will usually give me another two or three feet of depth into the material, unless I run into a large boulder that prevents my further passage downward. But if I do find a large boulder, it tells me that the material goes considerably deeper, and this is what I am trying to find out, anyway. If I do not run into a boulder, and if I still have not reached a contact zone, bedrock or run into good gold at the 6 1/2 to 7-foot level, I will pull out of the hole and go start a new one elsewhere.

I should mention that you have to be extra careful when doing the final coyote excavation as described above. By the time you get to this stage of the sample, there is usually a lot of material (rocks and boulders) hanging in the face of the excavation above you. You have to first make sure everything is safe enough, so that something does not fall in on you and cause an injury or worse. As every situation is different, safety in gold dredging must always be judged by the persons who are present on the scene.

There may be some experienced dredgers out there who do not entirely agree with this method. There are different ways to go about this; and even the best method in one location may not work very well somewhere else. This is something you will need to decide for yourself. The main point here, is that every dredger will have to decide how deep he or she will go into the streambed before giving up on a sample hole.

I am certain that I have missed a few deep deposits by pulling out of sample holes when I have decided that the material was too deep for my gear. But I am also comfortable that I have found a lot more shallow deposits for having covered so much more area by effective sampling.

If you do not have some system of deciding how deep you will take your own sample holes, it is possible to get over-extended with a single sample hole to the point where it can become a major production operation just to reach bedrock. But the whole idea behind sampling is to avoid getting into a major production operation until after you find a good-paying deposit.

As a good example of this, a couple of seasons ago there was a guy who came to ask my advice about his operation. He had a very large home-built 8-inch dredge downriver on his mining claim about four miles from where my base camp was located. There were five guys involved in his dredging program. They were hard workers, keeping the dredge operating during most of the daylight hours. But they were not recovering enough gold to even pay for the gasoline being consumed by the dredge.

In looking over the situation, I discovered that they had been at it full-time for over a month, but had not yet even touched down on bedrock! Their hole was 18 feet deep into loose material which kept sliding in on them, especially at night, partially filling their hole back in. Because the excavation was so large, all the cobbles and boulders had to be carried or winched out of their hole. This was a very slow process.

I advised them to move the dredge to a new location and sample for a shallower gold deposit. They were dead set against doing this; because they had already invested so much time, energy, and money into that single hole, that they could not allow it all to go to waste!

They also had a 3-inch dredge on the claim which they had brought along for their wives and kids to play with. I advised them to at least put one of their guys full-time on the 3-inch dredge, sampling around the claim to see if better and shallower ground could be located to place the production dredge. They thought this was a joke, all of them being certain that the 3-inch dredge was just a toy.

So they continued on in that same hole for the rest of the season. They finally reached a hard-packed layer of boulders at the 22-foot level (of material). The boulders were too large to be moved by hand and too tightly-packed to get a boulder harness around for winching. All they could do was spread out the hole in hopes to find an opening down to bedrock. They never made it. The first winter storm filled their entire hole back in with the same loose material. So there went an entire season, two dredges, five full-time operators, not a single completed sample hole, and not even enough gold to pay for the gasoline. Those guys didn't come back to dredge the next season, and we never saw them again.

***The whole idea behind sampling is to not allow yourself to get in too deep until you have verified a deposit that is worth the investment.***

How deep do you take a sample hole before pulling out because the area is too deep? This takes a bit of judgment, depending upon how good the prospect looks. You will have to decide.

I have run across a lot of gold dredgers over the years who considered prospecting in terms of a single sample hole. When they did not find high-grade in the location they chose, they figured that they had failed. This is the wrong approach. ***Prospecting is a process*** whereby you should think in terms of doing as many samples as necessary to follow the trail of small, single gold deposits along the path which gold winds itself down the waterway until you find those places where it really loads up.

## SAMPLING DIFFICULTIES AND HOW TO OVERCOME THEM

The hardest part of sampling is not the hard work involved with dredging new holes into the streambed, although this is certainly hard work. It is not in the substantial effort required to haul your dredging equipment into and out of the various new locations. Nor is it in having to work out new deals with the owners of the mining claims that you want to test. It is not even a combination of all these things that makes sampling tough to do. ***The hardest part of sampling is in not knowing where the pay-***

***streaks are located, or how long you are going to have to go before you locate one.*** The main problem in sampling is in overcoming the unknowns!

If you knew where the pay-streaks were surely located, negotiating the deals, moving the equipment around and dredging the sample holes would all be rather easy to do, just so you could start dredging up the gold!

In a sampling activity, you have to do lots of hard work to reach the different contact zones or bedrock, all in an effort to find gold which you are not certain is even there in the first place. Having to put out all this time and effort without any certainty that you will find anything valuable creates internal conflicts which must be overcome so you can keep moving forward on a sampling plan. The hardest part of sampling is in dealing with the internal uncertainties.

Starting a sample hole is far more difficult than continuing an excavation which has already been established down to bedrock. A lot of the time, when starting a sample, you don't even know how deep you will have to dredge a hole into the streambed to reach the contact zones or bedrock. This adds to your uncertainty. Sampling basically consists of starting new holes and taking them down to bedrock, or as far as your best-judgment allows you to go. And you continue to start new holes, and holes, until you find a pay-streak. None of this would be too difficult if you knew where the pay-streaks were, or if you were at least certain that you would find one. It is all the hard physical work piled on top of not knowing where the pay-streaks are, or if you will even find one at all, which makes sampling tough.

It gets especially tough if you have dredged many sample holes, having hauled your equipment into and out several different locations, and you still haven't located any pay-dirt.

I have seen many beginning dredgers complete a single sample hole and then give up because they did not find gold. I have seen others give up before they even finished the first sample hole; some who barely even got started on the first hole! ***In fact, I have seen people give up on sampling even before they got out of their vehicles to take a look!***

Sampling takes a lot more persistence than this. You have to be willing to stick with it. I am talking about the need for some emotional fortitude here.

It takes some hope of making a valuable discovery to drive the sampling effort. Hope is something emotional. When hope is lost, the sampling program is over. This is a personal matter between you and yourself. So if you are losing hope, it is time to pause and re-evaluate your whole program. If there is still hope, then it is time to get the work done.

I myself am not immune to this problem. When the going gets real tough, or I am prospecting a new area for which I have little or no background information to give me hope of success, I have my own doubts and uncertainties to overcome, just like anyone else.

The key to finding the gold deposits in a proven gold-bearing area is nothing more than continuing forward with your sampling plan, and following up on the positive signs when you discover them.

It gets easier once you have found enough pay-streaks to gain your own certainty that they are present in abundance, and that they can easily be found just by mechanically persisting with standard sampling procedure.

In a production activity, you will recover as much gold from the pay-streak as you are putting time and energy into it. So you tend to find yourself really pouring on the steam. And in doing so, you get paid handsomely, you are seeing an immediate exchange for all your hard work, you feel great about it, and you do not mind putting in the long, hard hours - because you are dredging up treasure and riches! Seeing all that gold being sucked up into the nozzle can keep you going beyond all normal endurance, especially when there are nice gold nuggets being uncovered as you are moving along. This can keep you pretty excited. Hard work comes easy when the rewards are immediate.

A production-mining activity can be somewhat accurately gauged by how much gold is being recovered. A sampling activity cannot be measured in this same way; and if done so, you will generally find yourself slowing down. When little or no quantities of gold are being recovered, incentive and momentum can be lost.

Probably the best way to measure how well that a sampling activity is doing, is by looking at how much ground is being tested by thorough and effective sampling.

If the pay-streaks are present, they will be located reasonably fast by a thorough and energetic sampling activity. If they are not present, you can move to a new area which shows better promise and continue on there.

So in sampling, if you put your frame of mind onto producing like mad to cover as much prime area as possible with good sample holes in all the right places, instead of putting your attention on all the gold that you are not getting, you will get much more done, find more pay-streaks, and manage to keep your morale up all along.

I should clarify this a little more: Effective sampling is not only about covering a big area with sample holes. It is more about searching for and following up on those clues which will lead you into the pay-streaks. It is a crime to do all the work involved with making an excavation in a likely area to find gold, and then not follow up on positive signs there because you are in a hurry to start a new sample somewhere else. Sometimes the first signs are just a small sniff of a scent, something you are not even sure is a sign at all. So you have to be an attentive and thorough investigator. One sample hole thoroughly done is better than five poorly-done samples. I am not talking about the size of the excavation. I am talking about really sniffing around to find where gold is traveling through the waterway. Be careful to not allow yourself to pre-judge or talk yourself out of taking a careful, honest look at what is present in each place that you sample.

Sampling is pure detective work. To do it right, you have to be willing to do it as thoroughly as possible, based upon your own best judgment of what needs to be done. You have to be energetic in following up those clues.

I can tell you story after story of how just the smallest signs led me into rich pay-streaks. I also have a fair number of stories to tell about pay-streaks I missed, but others found later, because I was not in the right frame of mind when the signs were right there in front of me.

Really, how well any of us do in prospecting always comes back to the individual decisions that we make. Each decision in a prospecting program is like a crossroads. Each decision sets your new direction. So it is important to be paying very close attention to what your options are. Prospectors who start a sampling program, or a sample hole, already knowing they will not find any gold there, often close off their perception of positive signs that are right out in the open.

I have watched prospectors who start many sample holes but never finish any of them. Even if they do a lot of work, they hardly ever find any gold. It is not that most of their holes were deep or difficult to do. They just didn't have the emotional endurance to complete their sample holes, deciding *"there is no gold"* after investing themselves into the excavations, but before giving them a fair chance.

Let me put it like this: If an area looks good enough for you to start a sample hole, than it is good enough to do an honest job of finishing the hole! Sometimes that just takes another 30 minutes or one more dive. You owe it to yourself to keep your personal hope generated until at least you have given each sample a fair chance. Otherwise, you will waste a lot of energy reaching out, but not far enough, or not thoroughly enough, to find what you are looking for.

One common way to make a sampling plan is to look the area over ahead of time and decide upon all the locations where sample holes should be dredged, and mechanically stick to that plan no matter what; unless through sampling, you discover that additional holes are needed to follow up on positive signs.

If you find yourself becoming too burnt-out to do a thorough job of it, it might be best to take a day or two off and then go back and hit it hard right where you left off. Probably the best thing to do, though, if you can, is to just push on through with it. Because the sooner you dredge the holes, the sooner you will find the pay-dirt, and the sooner things will be great again.

It has been said that intelligence can be defined as the ability to estimate the correct amount of effort required to overcome the barriers in completing a task. It takes so much effort to carry a filled gas container from the truck down to your dredge. It takes more effort to carry your dredge from the truck down to the river. If you use too little effort on a project, you probably will not complete the task. If you use too much effort at once, mistakes can be made - or at least some of the effort might be wasted.

In sampling, because so much physical activity is required, and because the main barriers to overcome are your own unknowns, there is no way that you can use too much effort to accomplish the task. Give it all you have got. This is the best way!

The pay-streaks are there and you will find them if you try hard and stick with it. And remember, as you continue to gain personal experience, it continues to get a lot easier.

# FALSE CERTAINTIES

After hearing me talk about some of the difficulties in sampling, you might get the idea that sampling is a long and drawn-out process. It can be, if you are not sensitive about following up on the positive signals, or if you have placed yourself into a lousy operating area. For the most part, though, it usually does not take very long to find the next pay-streak.

Once a pay-streak is found, you tend to get all excited and forget about all of the uncertainties that you felt during sampling. In fact, once you find a pay-streak, you can look back and see just how easy it was to find. After the pay-streak is found, the reasons why it is there become obvious. Every time I find a pay-streak, there seems to always be people around talking about how obvious it was that one would be there. I also do this when someone else finds something good. Hindsight is very clear in prospecting! If the presence of those pay-streaks was so obvious, how come someone else did not recover them before? The reason is because it is a lot harder to see where they are with certainty before you find them.

Once you get into a pay-streak, you can look back and see just how easy it really was for you to find the deposit. You can feel just great about your sampling abilities. You will consider how easy it is going to be for you to find the next deposit. You will be feeling at the top of your game. But don't let all of this fool you too much. This is the euphoria that is part of finding high-grade gold. When the pay-streak plays out and you have not gathered any new sure-thing hot leads, and it is time to go out and find a new pay-streak, all the unknowns kick back in again, and you will probably start worrying that you will never find another one which was as good as the pay-streak that just played out. This is the opposite side of the euphoria that is part of gold prospecting. It is normal.

If you are going to continue to play the dredging game, it is necessary for you to prepare yourself to handle these kinds of emotional extremes. One day you are not making any gold, and the next day you are pulling it out of the river by the handfull. Then after a while, the pay-streak will play out and you are not making anything again while you are looking around for a new pay-streak to dredge up. It is like a roller coaster.

It can be particularly-difficult to deal with when a long, rich pay-streak plays out. You will worry that you will never find another gold deposit as rich or better. I have seen a few dredgers quit after their first pay-streak played out, because they were not able to cope with this severe kind of change.

I have personally found that there is always a better pay-streak which can be found through more sampling, and there are always plenty to be found that are good enough.

# 3

# GOLD DREDGING PROCEDURE

## DEFINING ACCEPTABLE GROUND

It is a very good idea for you to define for yourself what amount of gold is minimally acceptable to you on a daily basis. It might be a half-ounce per day. It might be more. Or it might be less, depending upon the size of the dredging equipment you are using, your operating expenses, and how much gold you need to recover to make the effort worthwhile. The main point here is that you should set some minimum amount. If you are recovering that much gold on a daily basis, you will stick with it. If your daily averages start dropping below that point, you will start sampling around for better ground.

There are several good reasons for doing this. One is that if you do not know about how much gold is acceptable to you, and what is not acceptable, you can waste a lot of time *"going on hoping"* (for some undefined target) while you continue to dredge in low-grade material, rather than sample around for something better like you should. Another good reason is that you need to determine how far to each side and how far to the rear to dredge into the lower-grade material along the boundaries of a pay-streak.

Perhaps the best reason that you need to define a minimum acceptable level of daily gold for yourself, is so you can take in the better, and much better, and tremendously-better pay-dirt as a bonus when you find it. This could be on a regular basis if you become good at sampling. If you do not take on the good finds as a bonus, you can find yourself comparing the good finds with the acceptable finds; and pretty soon the acceptable finds may not be acceptable anymore! This is especially true when you are dredging up a high-grade pay-streak.

I know of a guy that was new to gold dredging, who was into a very large pay-streak which was paying him a little more than half-ounce per day for every day that he went out and dredged. He was happy with this, because he was making a respectable cash profit over top of expenses, and gold prices have been going up to make that even better. Everything was going along just fine until one day when he uncovered a bedrock up-cropping and pulled six ounces of gold out of a single pocket in the bedrock just in that one day. The following day, he was back into the half-ounce amounts again. Only that was no-longer good enough. He quit shortly thereafter. He left the area, never came back, and I have not heard of him since.

There will always be greater highs if you stick with gold dredging long enough to get good at sampling. But if the greater highs are the only thing that is now acceptable pay-dirt, you will find yourself frustrated most of the time.

If you define for yourself what is acceptable as pay-dirt, as you continue to get better at sampling, you will find that there is a surplus of acceptable ground available to you. Therefore, you will be able to upgrade your minimum acceptable levels a degree or two as time goes along.

If you are willing to dredge up every bit of acceptable ground that you can get your suction nozzle into, and are willing to accept the bonuses as they are uncovered, and just treat them as a bonus, you will be making more gold more of the time - and you will find more bonuses, too.

## DREDGING UP PAY-STREAKS

There are basically three different phases to dredging up a pay-streak - all of equal importance. They are, finding the pay-streak (sampling), locating the tail-end and outside left and right boundaries of the pay-streak, and dredging up the entire pay-streak.

One important thing to know about dredging up a pay-streak, is that you really only have one chance to recover the whole thing without having to go to a lot of additional effort. That chance is on the first try.

The first thing to do when an acceptable pay-streak is located, is to drop back and find its tail-end. Otherwise, throwing cobbles and dumping your dredge tailings on top of a portion of good-paying ground is <u>very</u> poor planning. It just means that you will have to eventually go back and move them all over again, if you are going to recover the gold from underneath your tailings during the same mining season.

Finding the tail-end of a pay-streak can be more than twice as hard to do if there is twice as much material sitting on top of the pay-streak because of your tailings. Dredge tailings are loose. When you are trying to dredge a hole down through them, they just keep on sliding down into your hole. During the night, water currents will just keep washing them into your dredge hole.

It is <u>much</u> easier to drop back right at the beginning, locate the tail-end, and start dredging from there. But, as much sense as this makes, you really have to discipline yourself to do it when you find a new pay-streak and start seeing all that new gold. It is <u>very</u> easy to become mesmerized by the gold, wanting to continue on so you can see more of it, to see how much is there, and to quickly dredge up as much of it as you can. The more gold you see, the more you want to see how much more is there, and so on it goes. This is the excitement of gold mining. Meanwhile, your tailings and cobbles are piling up on good, or perhaps even better, gold-bearing ground downstream from you.

I expect that most dredgers will have to dump a whole bunch of tailings on top of a rich pay-streak or two before you gain a full understanding of this for yourselves. I had to learn this lesson the hard way. Expensive lessons teach you to do things right!

Any time spent in finding the tail-end and the outside left and right boundaries of a pay-streak right at the beginning, will be very valuable time saved, and will help you recover more gold. The reason I say this, is that when you dump tailings onto a pay-streak, the additional work that it takes to remove the tailings later can render portions of the pay-streak unworthy – meaning that you can no-longer recover the minimum amount of daily gold that you have set for yourself.

Even if it is a very large pay-streak and it takes some time to discover the tail-end, it is usually worth your while to do it.

Sometimes, as you drop back towards the tail-end, you will notice that it is not paying as well as it was where you first found the pay-streak. This is very common, since you will want to drop back as far as you possibly can, as long as the pay-streak pays above your basic minimum acceptable daily production requirement. Dropping back can be hard to do if your minimum is a half-ounce per day, and you have already uncovered an area up ahead of you on the pay-streak that is paying ten or twenty times that much. Sometimes there a nuggets, too, which you are moving farther and farther away from as you drop back. It takes discipline! You have to overcome your own positive emotions. Easier said than done!

If you do not believe in gold fever, or feel that your personal discipline is strong as nails and cannot be broken, just wait until you have to drop back and find the low-grade tail-end of a pay-streak when you have located what appears to be extreme-riches up ahead of you!

One good way to look at it, is that your production results are certainly going to improve as you move forward on the pay-streak. This tends to spur you on quite a lot. You end up getting more done faster, just so you can get back up to where you were seeing richer volumes of gold. So you end up making more gold in the long run.

There are two different ways to find the tail-end of a pay-streak: (1) backtracking, and (2) dropping back while dredging holes. If the streambed material is relatively shallow over top of where the gold is located, and the water is not moving too fast in that section of river, one good way is to track the gold back to the tail-end of the pay-streak. This can be accomplished by placing your dredge slightly downstream and off to one side of where you found the gold, and by dredging backwards - meaning with the current, instead of against it. Follow the gold along until it plays out at the tail-end of the pay-streak (See figure 3-1).

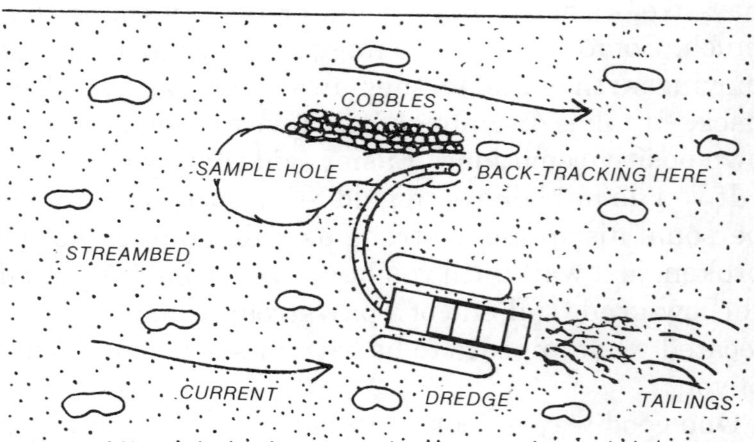


*Figure 3-1: Backtracking with the current in a pay-streak to find its tail-end.*

When *"backtracking"* along a pay-streak in this way, your main purpose is not to recover all the gold that you uncover. It is to move backwards as quickly as possible to find the tail-end of the pay-streak. You will recover all of the gold once you turn around and start moving forward again. Meanwhile, the more material that is disturbed while moving back along the pay-streak, the more misplaced cobbles and tailings that you will need to deal with once you do turn around. So ideally, backtracking is accomplished as quickly as possible, doing no more than is necessary to follow the line of gold back to where it is no longer paying well enough to make it worth your while.

The gold in almost every pay-streak will be found concentrated either on the bedrock or in the contact zone between two layers. For the most part, if the gold is in a pay-layer or on the bedrock in any acceptable amounts, you will be seeing plenty of gold while you dredge. At the head of a pay-streak, if the gold is being found in a specific layer, it is common to find it in the very same layer towards the tail-end of the pay-streak.

Sometimes the material changes and you cannot find the same layers as you drop further back on the pay-streak. In this case, it might be necessary to do a little more looking around to discover if the gold is there; and if so, where it is located now. This will give you something new to look for as you drop back even further. An example of this is where some change along the bottom of the waterway transfers the gold to the bedrock from a pay-layer or vice versa. I have seen this happen many times.

If the streambed material is too deep over top of a pay-streak to backtrack easily, or the water in that section of river is too fast for you to turn around and dredge with your back to the current, then instead of backtracking, you will have to use the other method of finding the tail-end of a pay-streak. That is to keep on dropping back, dredging new sample holes as you go along. This method consumes more time and energy than backtracking, because you cannot follow the gold by sight to it is tail-end. Instead, you have to guess where the tail-end might be, drop back to that point and dredge a hole while looking for that same area in the material that was paying further forward on the pay-streak. You have to look the area over closely to see if it is still paying, and how good. If it is still good, you have to discipline yourself to pull out of the hole, drop on back farther, and do it all over again. This can be really tough to do when you have to keep pulling out of good gold in front of you to start new holes behind you. But there really is no other way that I know of if you want to dredge up all the high-grade gold on one clean sweep.

Once in a while, you can find a pay-streak while sampling, drop back to find the tail-end, and not find any gold back there. Then, as you move forward, you find out that what you found in sampling is all there was - a single deposit! This happens, and can be kind of disappointing - but it is all part of the game.

When you are attempting to find the tail-end of a pay-streak, each time you drop back to dredge a new hole, it is wise to set up your dredge off to the side, where you estimate that the dredge tailings will be away from the pay-streak and out of your way when you come back up through.

Sometimes a pay-streak will be spotty towards the tail-end, or it may be a spotty pay-streak altogether. This is something to watch for while backtracking or dredging holes to find the tail-end. If you are backtracking a good showing of gold, and it suddenly ends, drop back even further and dredge a few quick holes just to see if you can pick it up again.

If you find a low-grade section of streambed material along the pay-streak, and the water is moving slow in that location, it might be a good place to start dropping your tailings so that you can start dredging forward towards the higher-grade material, even if the pay-streak might pick up again further downstream. Dredge tailings will usually just pile up in slow water, rather than travel downstream. So if they are dropped on a low-grade section of riverbed in slow water, they will usually give you no further trouble downstream (See figure 3-2).

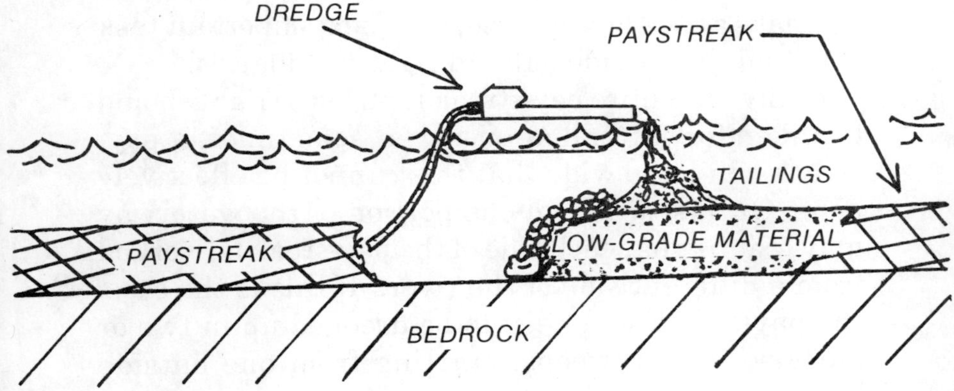


*Figure 3-2: Dropping tailings on low-grade material in slow-moving water.*

Fast water will sweep a fair portion of your tailings downriver, and will have them sitting in abundance on top of any pay-dirt that is located just downstream along the same line of where you are dredging. Since pay-streaks are almost always found along the very same gold path in the river, and sometimes quite close to each other, it is usually worthwhile in fast water to dredge a few more sample holes downstream to make sure that the acceptable pay-dirt does <u>not</u> pick up again where your tailings are going to end up. In the long run, it is best to know at the beginning what you are going to be dropping your tailings on top of.

All pay-streaks eventually play out. That is when it becomes necessary to go find the next one. A lot of the time, the next pay-streak will be found just upstream or just downstream of the pay-streak that you have already located. If it is just downstream, and you have just dumped tons and tons of tailings all over it, you are going to be very disappointed!

Once you have found the tail-end of the pay-streak and have a good hole started, your next important task is to find the outside left and right boundaries.

Ideally, the entire pay-streak from left to right should be taken forward all at once. But sometimes, a pay-streak can be so wide that this cannot be effectively done. Or, fast water in one portion of the waterway may require that each side of the pay-steak be worked from a different side of the river. If this is the case, the pay-streak will need to be dredged up in two or more separate sweeps, starting from one outside boundary and working towards the other as you move forward. When doing this, it is important to be extra careful to keep your tailings and cobbles well away from the non-worked section of pay-dirt, so they will not be in your way on the next pass (See figure 3-3).

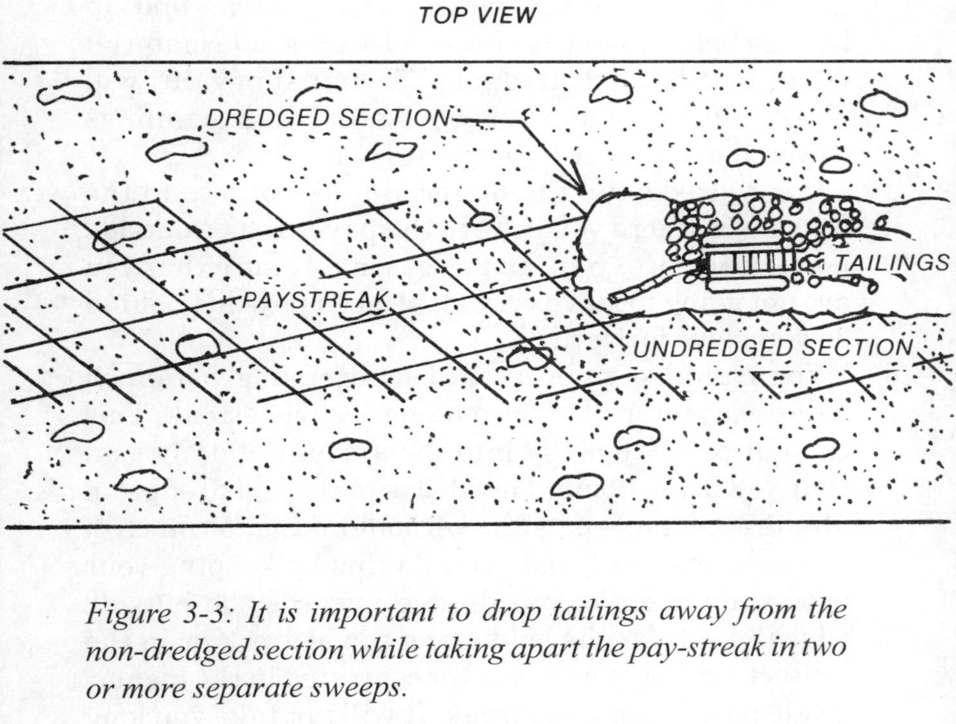


*Figure 3-3: It is important to drop tailings away from the non-dredged section while taking apart the pay-streak in two or more separate sweeps.*

One of the best ways to locate the outside boundary of a pay-streak is to get into gold inside the pay-streak and follow it outward to the side - just like in backtracking. If you know where the gold is concentrated in the material, and watch closely as you follow it out toward the side, you will see where it stops paying. This is often the outside boundary.

In that a pay-streak will sometimes pay very well along one part of the pay-streak, and pay less and less as you move further off to the sides, it is important to keep your minimum acceptable daily production requirement strongly in mind while taking apart the pay-streak. Otherwise, you can get so excited about uncovering the richest portion that you can end up leaving behind large portions of lesser-grade material which is still worth dredging. Then that pay-dirt will eventually lie under your cobbles and dredge tailings as you move forward (See figure 3-4).

If you dredge up the material a foot or two to the outside of where you believe the pay-streak's outside boundaries lies, you can be reasonably sure that you are not going to leave very much good pay-dirt behind as you move forward.

You can also go overboard by dredging too much valueless material to the outside of a pay-streak. This can cut pretty heavily into the amount of daily gold that you bring home. The idea is to try and dredge up all the acceptable pay-dirt without taking too much of the valueless material. You do this by keeping your eyes wide open as you dredge towards the outside edges. Look for the gold along the bedrock or in the contact zone, just as it was concentrated in the higher-grade part of the pay-streak. It will not take you long to see where the gold plays out. Then you dredge just slightly to the outside edge of that point as you move forward on the pay-streak. You have to watch closely as you dredge along that line for any signs that it might be improving (See figure 3-5). The outer boundaries of a pay-streak often change as you move forward.

As mentioned in Chapter 1, sometimes a specific line of heavier gold (larger pieces) will be found deposited along one outside edge of a pay-streak. This is

TOP VIEW

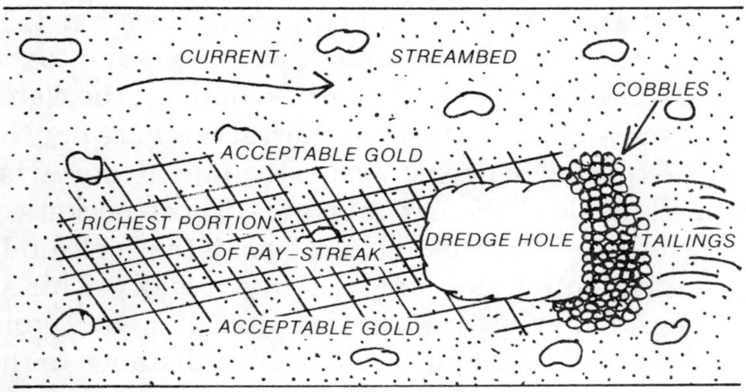


*Figure 3-4: The improper way of dredging just the richest portion of a pay-streak while leaving acceptable portions behind.*

TOP VIEW

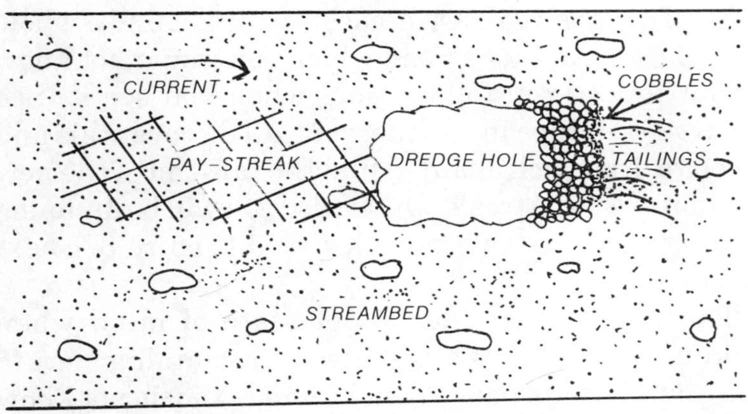


*Figure 3-5: The proper way of dredging just to the outside of acceptable gold in a pay-streak.*

something to keep your eyes open for as you dredge along the pay-streak boundaries. I have noticed that when this special line of heavy gold does exist, it usually lies along that boundary which is closer to the river's channel and faster water. Sometimes the water can be moving pretty fast out there, and there may be some exposed bedrock with streambed material trapped inside of the irregularities. Those pockets of material are worth looking into, even if they definitely appear to be outside of your pay-streak. This is because sometimes a special line of heavy gold (nuggets) will be found to run a short distance to the outside of the established pay-streak, as shown in figure 3-6.

Sometimes a pay-streak will be kind of spotty towards the outside boundaries. So you will have to watch even closer for where the gold shows itself - and follow it.

Always investigate immediately if you notice that the gold seems to be getting better towards the outside edge of a pay-streak. On occasion, you can find a whole new part of the pay-streak which you would not know about unless you follow gold when you see it (See figure 3-7). Once in a while, you will discover this and need to drop back again to find the tail-end of the new section of a pay-streak. Otherwise you will be dumping tailings on it as you move forward (Figure 3-7 next page).

Thoroughness is the proper state of mind when taking apart a pay-streak, always remaining alert to exactly what the pay-streak is doing. You do not know for sure unless you take a look.

You can go overboard on spending too much time looking, too. You can spend so much time looking that

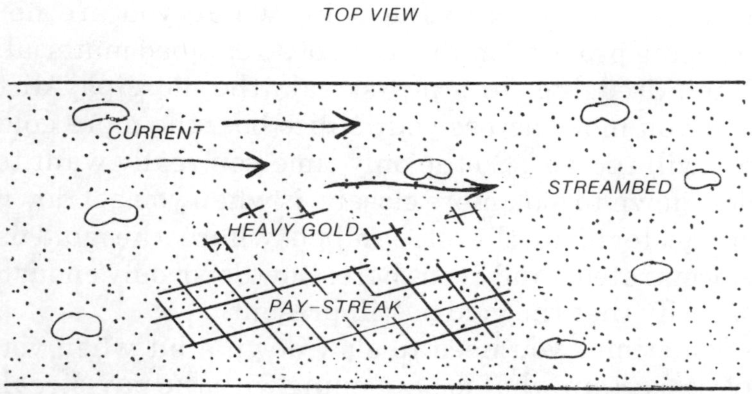


*Figure 3-6: Sometimes a line of heavy gold in the fast water can be found just to the outside of a pay-streak.*

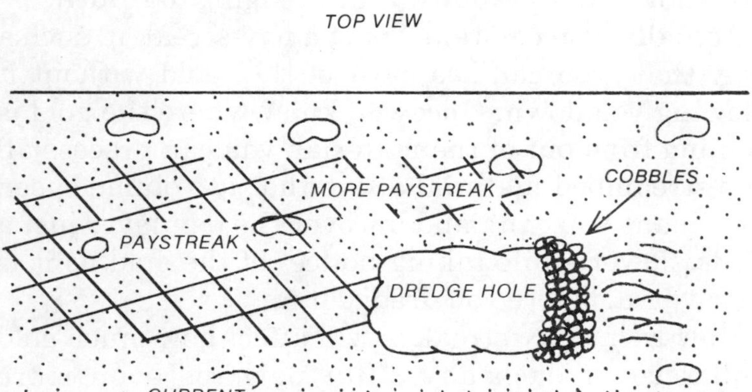


*Figure 3-7: Whole new sections of a pay-streak can be found by following the signs of gold.*

it slows you down to the point where you are not dredging production-amounts of streambed material. When dredging in a pay-streak, the more pay-dirt that you move across your sluice box, the more gold you will recover. So the only time you really want to slow down to look more closely, is when you get down to the level in the streambed where the gold is concentrated. And even there, slow down only enough to verify that the gold is still present.

Sometimes you need to slow down a bit when you are uncertain what is happening with the pay-streak - like when establishing exactly where an outside boundary is located. There is no need to slow down too much when you are in the meat of a pay-streak and know it. You will have plenty of time to admire all that gold later, and there will be more of it, if you do not allow it to slow down your dredging too much.

Actually, you can take apart a pay-streak in such a way that you can see most of the gold without it slowing you down. Once you know where the gold is coming from out of the material, you can process all the streambed material above that at volume speed (as fast as you can), and uncover the pay-layer just a little slower while taking notice of the gold as it is uncovered, before you dredge it up.

How rich a pay-streak is will affect how much gold you recover each day. But perhaps even more importantly, how much volume of streambed you are able to process will directly affect how much gold you recover – and how quickly and often you will find pay-streaks in the first place. Remember this: ***Your ultimate success in suction dredging comes***

***back to how much volume of hard-pack you can direct up the suction nozzle.***

I knew a dredger a few years ago that was so hooked on the positive emotional charge he experienced from uncovering gold underwater, that he dredged everything at slow-speed, even the gravel up in the streambed where there were no gold concentrations. He did not want to miss seeing a single piece of gold when it was uncovered. He was also complaining nearly all the time that he was not getting very much gold.

This was a really smart guy; he was a retired engineer from NASA. I had quite a few discussions with him. He knew that gold is extremely heavy and that it concentrates either on the bedrock or in the contact zone between flood layers. But he was so addicted to the feelings he experienced when seeing a piece of gold underwater, that he could not keep himself from looking for pieces of gold everywhere, even where it was very unlikely to be. He never did get much gold, because he never developed the skill to process volume amounts of streambed. The more gold he saw down at the pay-layer, the more he slowed down everywhere else. I encourage you to not make the same mistake.

Sometimes the streambed material will change in color, depth, consistency or hardness towards the outside boundary of a pay-streak. When this happens, do not assume that the gold ends at the same place where the material changes. Sometimes it does. Sometimes it does not. You have to look and see.

Sometimes you run across shattered or decomposed bedrock which can be saturated with fine-gold, medium-sized gold or even nuggets down inside the cracks - and way down inside of the bedrock. In this case, if doing so meets the minimum daily production requirements that you have set for yourself, you will want to break open the bedrock and get into it as deeply as it pays to do so. If the gold is there in acceptable quantities, you will be seeing it as you break open the bedrock.

Sometimes working the bedrock can really slow down your forward-moving progress on the pay-streak; and you will have to force yourself to do a thorough job of it because of the better pay that rests on top of the bedrock further forward on the pay-streak. You have to do it right, though. The only other alternative is to leave the gold behind underneath your pile of cobbles and dredge tailings, something you will inevitably regret having done once the pay-streak plays out.

I have seen bedrock in some places pay very well in fine-gold as far as three feet below the normal bedrock surface. It was shattered bedrock, all full of cracks that were underneath a fine-gold pay-streak which rested on top of the bedrock. It was obvious that the bedrock was paying well down inside, because I could see the gold as I broke apart the cracks.

Some bedrock does not pay very well, even when all cracked up or decomposed. In this case, you will not see the gold, and it will not end up in your sluice box.

Be careful to not allow yourself to get so slowed down trying to get just a few small gold flakes you see trapped in the bedrock, that they prevent you from reaching your minimum requirements. You will have to use your own best judgment where to invest your energy, so that you do not leave too much gold behind, but still move forward well enough to make the overall program worth your while.

## SPOTTY PAY-STREAKS

Some pay-streaks pay rather consistently all throughout. These are very easy to dredge. Other pay-streaks are spotty all throughout, so they can be more difficult to follow along. A spotty pay-streak tends to pay good, then poor, and then good again, and so on - not just as you move forward, but from side to side, too. So it is more difficult to follow a spotty pay-streak and keep track of the outside boundaries (See figure 3-8).

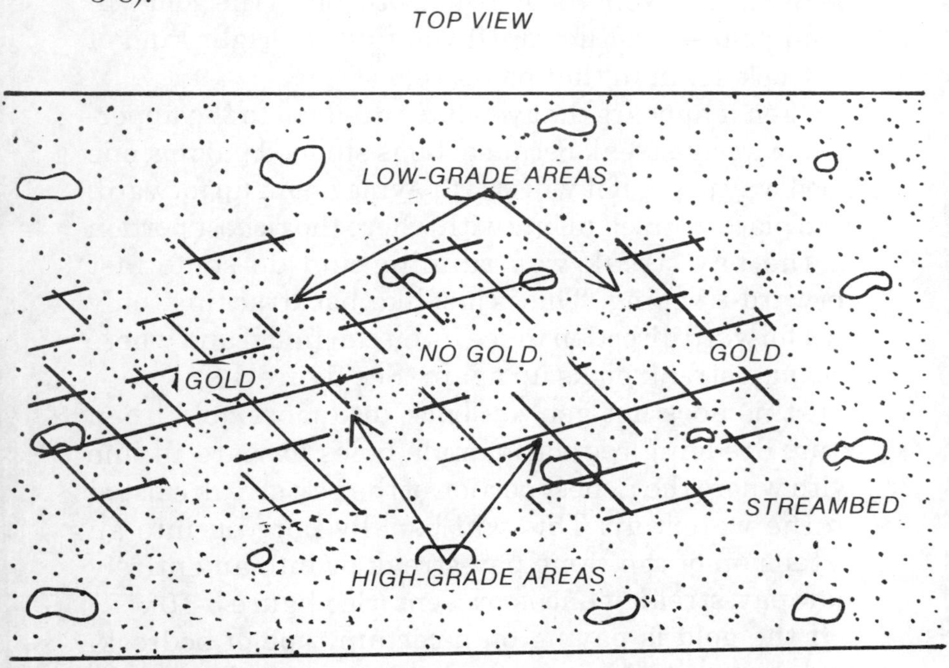


*Figure 3-8: Spotty pay-streaks are more difficult to follow.*

The only way to find and recover all the gold out of a spotty pay-streak is to dredge the whole thing up, taking the good with the bad. This can be done if the daily averages meet your minimum requirements. If they do not, then you will have to sample around for the richer portions and take them as they are found. This is relatively easy to do in shallow material, but becomes increasingly more difficult as you must penetrate deeper streambed material with your sampling.

One of the main reasons why a spotty pay-streak will go hot and cold is because of changes in the bedrock. If you are dredging along on a bedrock pay-streak (gold lies on and inside of the bedrock), and suddenly the bedrock goes from rough to smooth, often the gold will disappear altogether until you find the right kind of bedrock again in that pay-streak.

When it appears that you have dredged to the upper-end of a pay-streak because it has stopped paying, one good way to see if it will start paying again up forward, is to place yourself in line with where the richest portion of the pay-streak was running and do some fast forward-tracking. This is just like backtracking, only in a forward direction to see if you can find if and where the pay-streak picks up again (See figure 3-9).

If this does not get you back into more gold, then start dredging regular sample holes forward in line with where the richest portion of the pay-streak existed in the waterway. This will usually put you into an extension of the same pay-streak or into an entirely new pay-streak if one is present (See figure 3-10).

If the gold is paying on a certain type of bedrock, and then quits because of a change to smooth bedrock, as you move forward with sample holes, you are

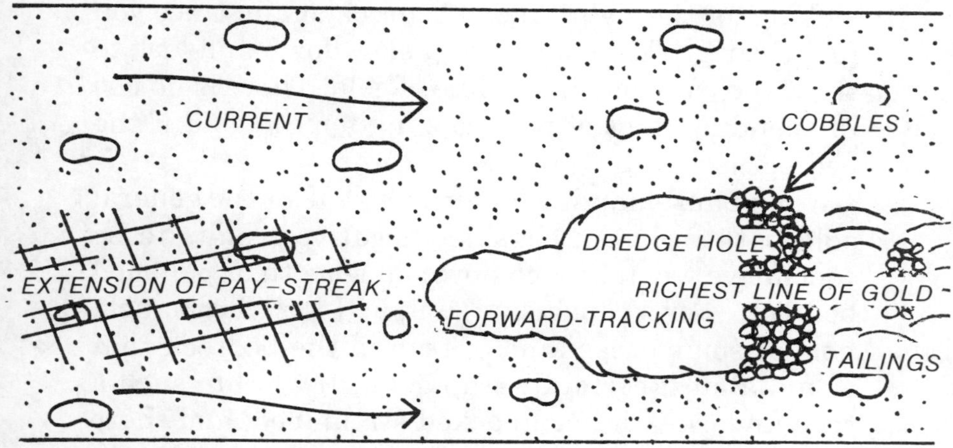


*Figure 3-9: Forward-tracking to find an extension of the pay-streak*

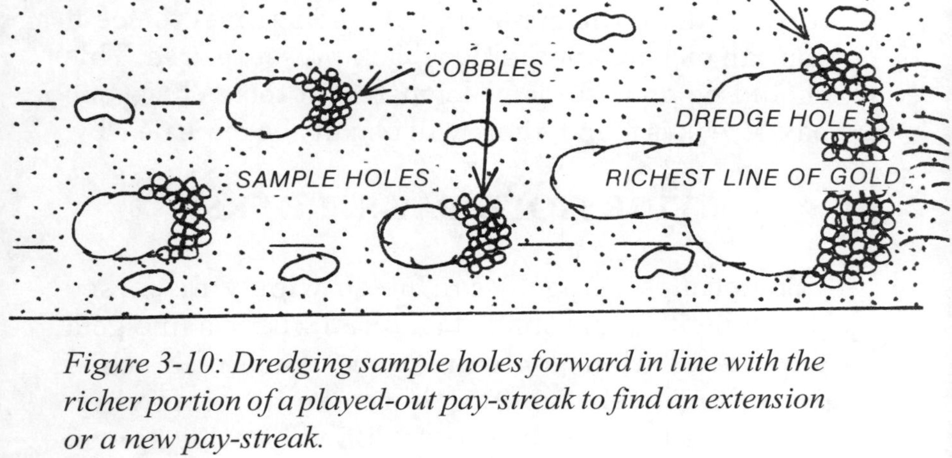


*Figure 3-10: Dredging sample holes forward in line with the richer portion of a played-out pay-streak to find an extension or a new pay-streak.*

looking for the right kind of bedrock again. Once you find the right kind of bedrock, slow down and look to see if you can pick up the line of gold. There is no use in spending a lot of time looking for the gold if the bedrock is wrong.

When I am sampling for a pay-streak or the general gold path, I do not consider negative results to be conclusive when I touch down on smooth or polished bedrock. For a conclusive sample, I look for a crack, hole or some other kind of trap in the bedrock that would almost certainly trap gold. If I locate such a trap underneath hard-packed streambed, and there is no gold in it, I usually conclude that I am either off the gold path or located in a place where water forces during flood storms are too great to even allow gold to stop there. If I do not find any gold in a good bedrock trap underneath hard-pack, I almost always immediately pull out of the sample hole and move to the next place that deserves a sample.

If you cannot find an extension of the same pay-streak just upstream, start sampling further and further along the same line of gold's travel to see if you can pick up another low/back-pressure area. This should be done further downstream too. Where one pay-streak is found, there will usually be more nearby.

## FINE GOLD PAY-STREAKS

Sometimes the gold can be so widely dispersed throughout the streambed material within a fine-gold pay-streak, that you cannot see much of it while dredging, even though there might be paying-quantities present. It is more difficult in this kind of pay-streak to tell where the outside boundaries are located. Pretty much, the only way to accurately test the pay-dirt in this type of pay-streak is to dredge up a fair-sized portion of it and then look into your sluice box to see what kind of increase in gold has been accumulated. Sometimes you even have to clean-up your sluice to tell for sure. In other words, it can be necessary to do short production runs towards the outside boundary to see if you recover enough gold in that area to make the effort worthwhile.

When I am into a pay-streak of this kind, I will sometimes spend the first 30 minutes or so of each production-day dredging a portion of streambed along that boundary which I am most uncertain about. Then I will look into my sluice box. Since I clean up the high-grade portion of my sluice at the end of each day, any new visible accumulation of gold will have come out of that first section of streambed that I dredged up. Sometimes it is necessary to pan down the high-grade section of my sluice to see how well I did on my short production run. In this way, I can estimate about where the outside boundaries of the pay-streak are located.

When I am into a rich fine-gold pay-streak, and notice the sluice box accumulating a lot of gold values, I will normally dump the whole sluice off into a wash tub, continue on dredging, and then dredge those concentrates back up again at the end of the day, before cleaning up. It seems wise to get an accumulation of fine gold out of a dredge sluice box every once in a while. They will be safer in a wash tub!

When you dredge up a set of sluice box concentrates like this, be sure to suck them up slowly - with your dredge operating at normal production speed. This is because if you dredge them up with your dredge running at a slower speed, there might not be enough suction-power to pull the heavy minerals and gold up through your suction hose. If not, the concentrates will load up in the hose; and when you finally do turn up your dredge, the concentrates will all be sucked up into your sluice box at once. This will overload the riffles and you will lose gold as a result. Riffles in a dredge sluice are pretty-much designed to concentrate well when the dredge is operating at normal speed. Generally, you will find that the riffles will load up with heavy concentrates if you do not have enough water flow passing over top of them

## WASTING GOLD

Sometimes you can start believing that fine-gold and lower-grade material is not that valuable, when you compare it to the richer portions of a pay-streak and/or the nuggets when they are present. Gold is an extremely valuable commodity, so it does not take very much of it to amount to some value. For example, at today's rate of exchange, one grain of gold is worth more than a dollar. This means that a fourth of a grain would be worth more than a quarter in today's buying-power. It does not take much of a gold flake to amount to a fourth of a grain in weight. Or how about a tenth of a grain being worth more than a dime? Yet, because flakes of that size are so small, you can find yourself easily overlooking them (discarding the value) - even when they are abundantly present in the streambed material. If all those flakes could be magically turned into nickels, dimes, quarters and dollar bills, I'll bet a lot fewer of them would be left behind by dredgers who are more eager to get at the richest portions of the pay-streaks!

The truth is that those small and tiny flakes of gold have far more intrinsic value than today's coins do, and this is going to be more so as time goes on.

So just as you must dial yourself into looking for the smallest signals when sampling for a pay-streak in the first place, it is just as important to dial yourself in when mining a pay-streak to make sure that you do not leave valuable portions behind. This is more difficult, because it is easy to get mesmerized into following only the richest portion of the high-grade when it is present.

## *"IT HAS ALREADY BEEN DREDGED!"*

Just about every good gold-bearing general area has specific locations or stretches of river that were once very rich, but are reported to have already been dredged out by others who were there before you. It is seldom true that these locations and stretches of river have been thoroughly dredged, even though the locals may believe they have.

A lot of the time, in the *"rush"* to get at the richest sections of pay-dirt, cobbles and tailings are dropped on top of the richest sections and other high-grade portions of pay-dirt. This is even more likely to happen when more than just a few miners were in there mining the location at once.

It takes years and years to thoroughly dredge out even a very short stretch of rich-paying river, unless the streambed material is very shallow and there is not very much of it. If the entire river was not dredged from bank to bank, then some portions were left behind. Since those sections of streambed were not dredged up, who knows for sure what amounts of gold are down inside of them? Those portions might be rich, or even richer, than the portions which were already mined.

Up until this point, I personally have only run across a small percentage of dredgers who know how to, and who are willing to, thoroughly clean-up a pay-streak. There are even fewer yet who know how to find the additional pay-streaks that lie along the same common line of travel in the river. It is always possible that the dredgers who reportedly cleaned out those extremely rich sections of river were not fully in the know about these things, and so left some excellent ground behind. In fact, it is likely that they did!

Also, when multiple people are mining in an area over a period of years, they tend to lose track of who dredged what and where. When tailings are dropped on a section of river, it is not likely that anyone is going to look underneath those tailings, especially when rich gold is being uncovered further upstream. Winter storms usually blow most of the tailings out of there. But by that time, everybody figures the area has already been cleaned up.

A lot of this sort of thing happens when a number of dredgers are working in a rich stretch of river. Everyone is in a *"rush"* to get at the best pay-dirt before someone else does. This especially happens in active mining clubs and associations when lots of dredgers are working a rich stretch of river.

When I see an inexperienced dredger find nice pay-dirt, I always take note of where the person's production hole was started. Nine times out of ten, the dredger will not drop back to find the tail-end. Nine times out of ten, the person will not go back and finish what he missed, because his or her tailings are then in the way! Inexperienced dredgers usually leave just as much gold - or more - behind them as what they dredge up. I wasted my own fair share during my early days, too. I have cleaned up after a few guys since.

When you hear that a very hot stretch of river has already been dredged out, the guy who is telling you probably believes it, even if it is not true.

When going in to check out such areas, there are ways to find out which portions of streambed have already been dredged, and which portions have not been dredged. Since hard-pack is created by major flood storms, if it is present in the river, obviously that location has not been dredged (since the most recent gold deposits may have formed in that very same area).

Winter storms usually fill in earlier dredge holes with loose material. Sometimes, you can look at the river from just the right angle, and you can see where the non-dredged portions of natural streambed stand out from the less-firm dredged places. When they are sitting right next to each other, the surface of loose material looks different than the surface of undisturbed portions of natural streambed. Sometimes there is a slight indentation where the earlier dredging took place. On some waterways, the difference will remain obvious even years after the holes were dredged and filled back in by winter storms (See figure 3-11).

Unless it is located in a very fast-water section of river, dredged material will not return to a natural hard-packed streambed formation until the next major storm comes along - which will redeposit the whole area, anyway. Until that happens, there is often a visible difference from the surface which the dredger can look for when searching for the non-dredged portions of streambed within a heavily-worked stretch of river.

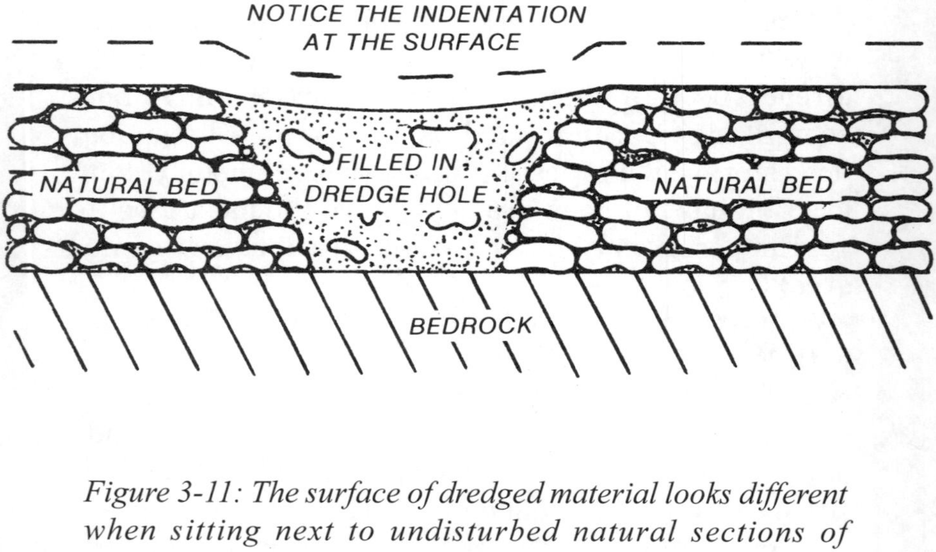


*Figure 3-11: The surface of dredged material looks different when sitting next to undisturbed natural sections of streambed.*

If the non-dredged portions of streambed are not easily recognizable by sight, you can put on your face

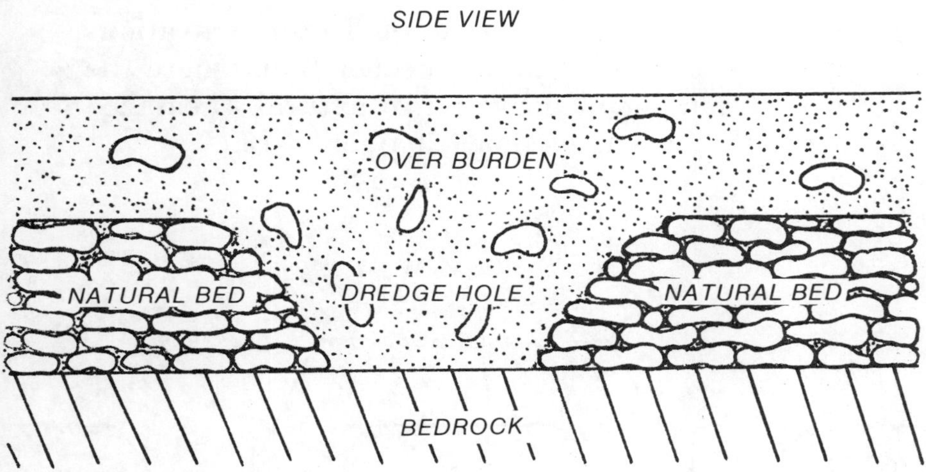


*Figure 3-12: Loose overburden can hide signs of non-dredged sections of streambed.*

mask, swim around, and pull cobbles out of the streambed from different locations here and there. Natural streambed should be much more hard-packed than the dredged material. When cobbles are pulled loose from hard-pack, a deep-colored silt is likely to be released. A filled-in dredge hole usually will not do this. The color and manner in which a natural, hard-packed bed is held together should be quite different than the dredged portions of streambed.

Sometimes, there can also be a layer of loose overburden resting on top of the dredged and non-dredged sections of streambed, as shown in figure 3-12. This can make your job of finding the non-dredged portions more difficult; because the overburden, even if shallow, is likely to obscure surface detection by sight or by pulling rocks from the streambed at the surface. Your best bet in this case, is to dredge a systematic set of sample holes. These excavations should not have to go any deeper than where the overburden should leave off to natural streambed when it is present, as shown in figure 3-13.

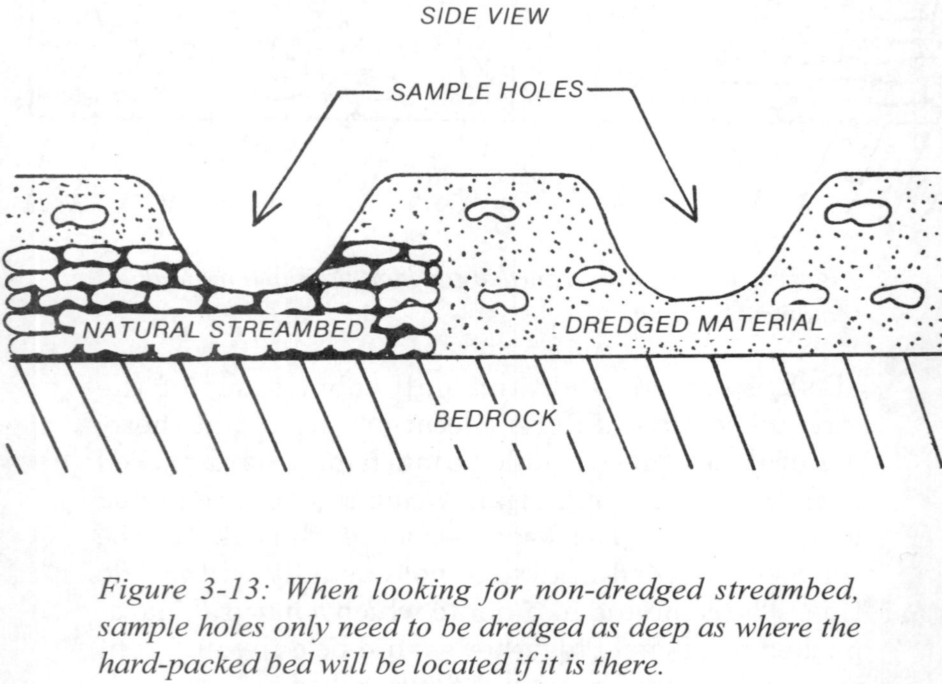


*Figure 3-13: When looking for non-dredged streambed, sample holes only need to be dredged as deep as where the hard-packed bed will be located if it is there.*

When looking for non-dredged portions of natural streambed in this way, if you are uncertain what the natural portions look like in that area, you might want to find some in a place which has certainly not been dredged - like off to the side somewhere. Then, when you start dredging holes, you will know exactly what you are looking for.

By the way, one of the best areas to find non-worked pay-dirt in the *"already dredged out"* stretches of river is in the fast water.

## DREDGE SIZES AND SUCTION HOSE

Under most conditions, in experienced hands, a larger-sized dredge (hose diameter) will move substantially more material than a smaller-sized dredge. For example, an experienced dredger can move about twice as much material through a 5-inch dredge as he or she can with a 4-inch dredge under similar conditions. You should be able to move around twice as much material through a 6-inch dredge as you can with a 5-inch dredge. An 8-inch dredge can process about twice as much material as a 6-inch dredge in most situations. This mostly has to do with the size of rocks that can be sucked up the hose. If bigger-sized rocks can be sucked up, fewer rocks need to be packed out of the dredge excavation to get the same amount of work accomplished. This makes things go much faster.

Larger-diameter suction hoses are generally also thicker-walled, heavier and less flexible than smaller-diameter suction hoses made of the same material. Therefore, while a 5-inch dredge can move twice the volume of material, it also requires more effort to manage a 5-inch hose, compared to a 4-inch hose. This same thing holds true as you move on up to larger-sized suction hoses. The larger they are, the more effort they require to manage well.

Some guys are strong enough that they can operate an 8-inch suction hose effectively by themselves. Most people are not. This is not only about the size of the suction hose. It is largely about the depth of the streambed material. No matter what size dredge is used, all oversized rocks still must be moved away from the working face of the excavation. Since a larger-sized dredge hose takes more effort to manage, deeper excavations almost require an underwater assistant to help move oversized material out of the way of the suction nozzle as progress is made.

I am referring to the types of hoses that usually come with store-bought suction dredges, and which are generally available from most mining equipment shops. Thinner-walled hoses are available if you look around. These will be lighter and more flexible, but that does not necessarily mean they are easier to dredge with. A certain amount of rigidity is helpful, so that you can pick the hose up off the bottom of the waterway and move it forward and backwards as you work the face of your excavation. If a suction hose is too flexible, it tends to get around and inbetween different rocks along the river bottom, and especially get caught up in the cobbles and boulders that stack up behind you. So if a hose is too flexible, rather than struggle with stiffness, you are constantly having to go back and free it from obstructions along the bottom.

There is a balance between maintaining enough rigidity in the hose that it is not continuously caught up along the river bottom, while not being so thick that you have little or no flexibility. You will find that most of the store-bought dredges available on today's market have found the proper balance.

Larger-sized dredges will not speed things up under all conditions. A 5-inch dredge will go into a lot of places that a 6 or 8-inch dredge will not go. I am talking about fast and/or shallow water. Some places have so many boulders in the streambed, that it makes little difference what size dredge you use – because most of the progress is made through winching. Bedrock can be so rough and irregular in some places, that you cannot get an 8-inch nozzle into the cracks and crevices far enough to recover a lot of the gold. Sometimes you can actually be more productive with a smaller-sized dredge. Each situation is different.

The key to operating a larger-sized suction hose effectively is in positioning the dredge so that you do not need to do any more fighting with the hose than is necessary. You may have to move the dredge forward more often this way, but you will also get more done with less energy. Otherwise, you will be fighting the hose most of the time. This can really wear you out.

If the water is deep enough, a larger-sized suction hose can often be suspended off the bottom of the waterway by hanging it from several floats. This helps a lot! Although faster-water conditions usually require the suction hose to rest on the bottom of the waterway.

One good way to position a smaller-sized hose, is balanced over one thigh. This way, you can operate the nozzle with one hand, while tossing cobbles with the other, as shown in figure 3-14.

## FAST WATER

It is often necessary to wear more weight while dredging in fast water. Even if this is the case, the fewer belts you use, the better off you are if you need to get rid of the weight in an emergency. One of the main reasons for this, is that when you are dredging with more than one weight belt, they tend to shift and rotate around so that it is more difficult to find the quick release catches. This is an important concern when dredging in fast water or in other situations where there is less margin of safety.

You can get more weight onto a single belt if you use the 15-pound weights, rather than the smaller-sized ones. Heavier weights tend to shift around on the belt, but this can be resolved by drilling holes and bolting the weights onto the belt so they will not slide, as shown in figure 3-15. Short concrete nails (with washers) can also be used to attach lead weights to a web belt once you position them where you want them to be.

Dredging in swift water can be very dangerous. I do not recommend doing it unless you are really up to it and have some experience under your belt. Safety in dredging is a personal matter. Each person has to make his or her own judgments about what can be safely accomplished and how to go about doing it.

When dredging in swift water, it is important to be especially careful how you connect things to your body. Never use ropes underwater, because they tend to get tangled around just bout everything, and can prevent you from making a hasty escape if it becomes necessary.

As working fast water is nearly certain to require the use of hookah, you should give some careful thought to how you will connect the airline to your body. In my own operations, we do not use the safety web harnesses that are sometimes supplied with hookah air systems. This is because a harness adds

*Figure 3-14: Balancing the suction hose over one leg so it can be operated with one hand. Your other hand can then be used to move cobbles.*

another quick-connect buckle that must be released to free yourself in an emergency. This might not sound like much in the comfort of this book, but having to get out of another encumbrance during an underwater emergency, when you already have one or two weight belts to get off your body, could mean the difference between life and death.

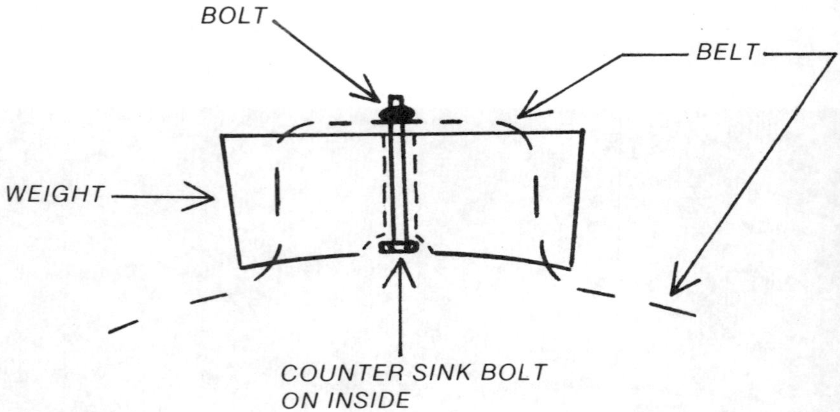


*Figure 3-15: Larger-sized lead weights can be bolted to the belt to keep them from sliding around.*

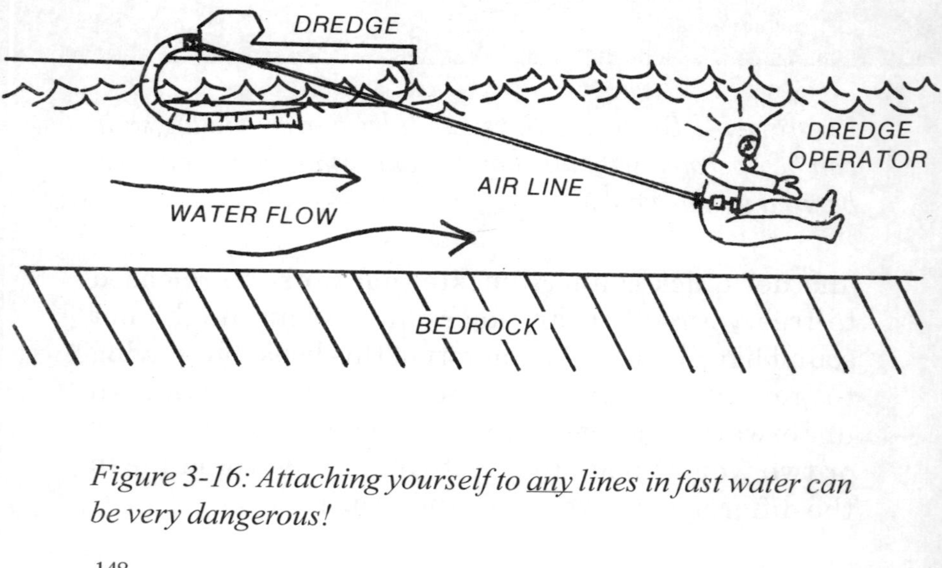


*Figure 3-16: Attaching yourself to <u>any</u> lines in fast water can be very dangerous!*

*The best pay-streaks can often be found under the fast water!*

Rather than use a harness, we tie a simple knot in the airline and connect that to our main weight belt. There still remains some amount of danger in this system; because if we get swept out of our dredge hole in fast water, we may find ourselves dangling at the end of an airline. But at least if we need to bail out, by releasing the weight belt, we also release ourselves from the fixed connection of the airline.

When dredging in swift water, your suction hose can become swept out of your hole and down river to drag

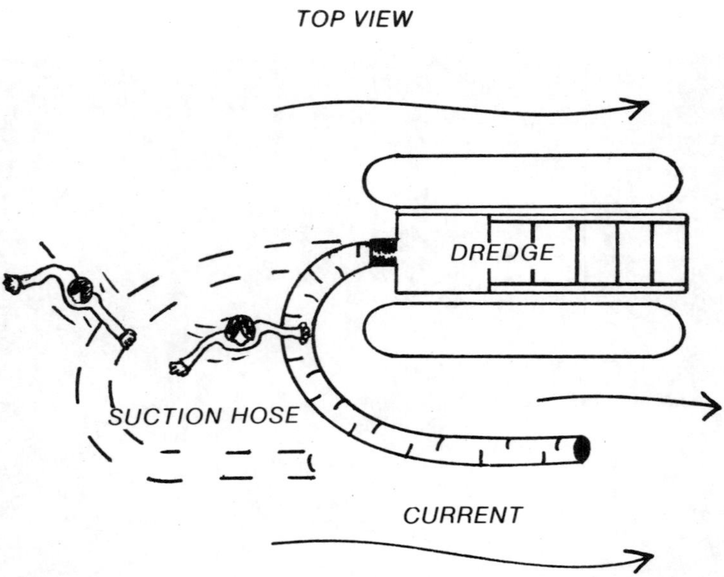


*Figure 3-17: Working the suction hose forward from the power-jet.*

from the power-jet. If this happens, the easiest way of pulling the hose back forward again is by starting at the jet and working it forward, as shown in figure 3-17. This takes much less time and energy than going back to pull the nozzle forward.

## FAST WATER HARNESS

When dredging in moderately-swift water, there is a harness that you can easily rig up which will keep you from losing very much of your suction hose if it does get swept out of your hole by the current. This harness also makes it much easier to quickly pull your

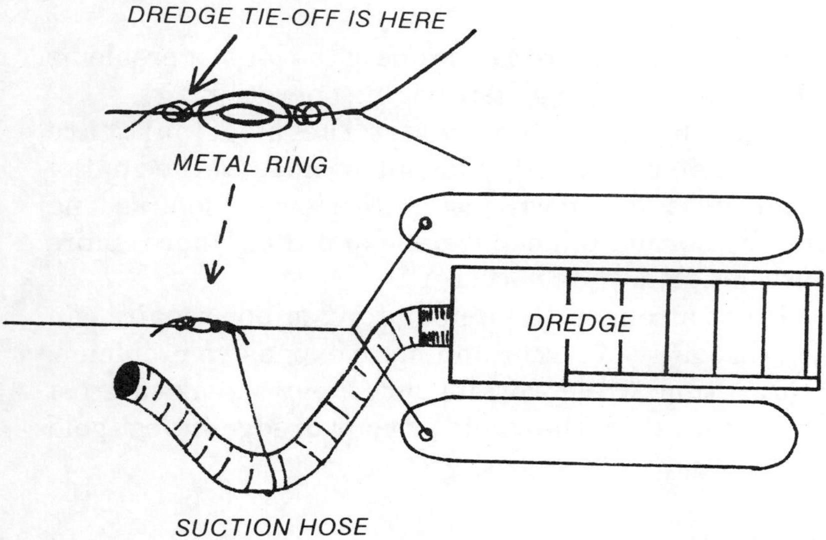


*Figure 3-18: Fast water harness.*

dredge further upriver when you need to. The harness is rigged so the dredge tie-off is located right above where you are working the suction nozzle. This way, when it is time to move the dredge forward, you can do that and pull the hose forward at the same time (See figure 3-18).

After you have set up your dredge in a new spot along the river and have started dredging again, it is a good idea to look back every once in awhile to make sure the dredge is riding well. This is especially needed when you are dredging in fast water.

# MAINTENANCE AND REPAIRS

Equipment failure can be one of the largest problems and expenses to a gold dredging operation.

Store-bought dredges, for the most part, are dependable, have all of the major bugs worked out of them already and will last quite a few seasons as long as the normal upkeep is done and the minor repairs are handled when necessary.

If not attended to immediately, minor repairs can be the cause of larger and more expensive problems. Down-time while working a rich pay-streak can cost you more than the value of your dredge in lost gold production.

# 4

# DEALING WITH GOLD, CLAIMS AND PARTNERS

I once knew a very knowledgeable guy who had spent a considerable amount of time studying the characteristics of gold. One of his conclusions was that each different kind of matter (the basic elements) emits a different type of wavelength; and that the wavelength emitted by gold is a wavelength of beauty in the aesthetic band. That gold is a very pure and beautiful substance is not in question.

But wait a minute. Beauty and value are perceptions or feelings that are experienced by human beings, not by a material substance. A person feels that gold is rich and beautiful when he looks at it closely and/or comes into contact with it. So we are discussing here the way people commonly feel when they have some experience with gold – or sometimes just the concept of gold. These common reactions are: (1) admiration of the pure, aesthetic quality and richness of the substance; and, (2) the desire to have some for yourself;

and, (3) feelings of euphoria when you actually acquire some of your own gold, especially when you have made the discovery in some hidden place; and, (4) the fear or worry that someone might try and take your coveted prize away from you, or that you will not receive all of what is rightfully yours; and, (5) the impulse to take necessary steps to protect your interest from others.

Please understand that I am not saying that any of this is wrong. I am just saying that these are the normal sequence of emotional reactions that play themselves out when human beings search for and discover high-grade gold deposits. Add high-grade gold into the never-ending challenge of just normal human relations, and you are potentially in for a lot of drama, unless the various relationships are structured thoughtfully and managed carefully.

Gold is a very rich and alluring substance to deal with. No-matter how much gold a person has, there never does seem to be quite enough of it to completely fulfill one's desire for more. Very much like an addictive drug, the more gold that you have, the more you realize that it is not nearly enough to fully satisfy your desires.

Perhaps gold does not affect everyone like this. But it certainly strikes most gold prospectors this way! To those prospectors out there who would disagree with this, my answer is that you just have not found enough gold yet to get your juices flowing!

Over the years, I have heard quite a few gold prospectors declare that they are immune to all of this gold fever nonsense, saying they are just into dredging for the money and economic reasons. Some say they are into it just for the fun and adventure – or for something interesting to do. But I say that you really cannot be certain of your immunity or how you will react until you uncover a rich gold deposit of your own. Usually, it is the nonbelievers who lose themselves the worst once substantial amounts of raw gold come into play in a gold mining project. My own conclusion is that no one is immune. It is just a matter of how much gold that it takes to draw you out. Usually it does not take very much!

But even if you are one of the few people on this planet who are entirely free of the emotional impact from large quantities of raw, pure gold (very doubtful, or you probably would not be reading this book), if you are going to get involved with a gold mining program, you better plan on the effect that gold will have on others!

Let me first start by explaining that I am not a licensed attorney, so I am not able to give out legal advice. But I am an experienced prospector and miner, and have had the opportunity to become involved, or stand on the sidelines and watch, many, many prospecting and mining ventures make a go at it. The views expressed in this chapter are not meant to give you legal advice. You should seek out a licensed attorney for that. My purpose here is to point out some of the different situations and relationships that you may encounter by circumstance as you move forward in a gold prospecting program of your own. How you ultimately deal with the different situations is your own business. My views expressed here are expressed only to give you a look in advance at the types of relationships that you may encounter while gold prospecting.

It is very important to be careful about who you partner-up with when dealing with any amount of raw gold, especially when you are planning a venture with the intention of dredging up large amounts of gold out of a river.

When negotiating the various dealings which you will have with partners and/or the owners of mining property, it is very important to be extra careful to make deals which all parties are agreeable to, and which all parties will be able to stick to. Agreements should be made very clear, in writing if possible, and always before the first high-grade gold deposit is located!

Ultimately, the structure of an agreement should be put into place and finalized as the next step beyond discussion of the idea between the partners – before you begin organizing or capitalizing the project.

The process of putting agreements in writing forces the persons involved to address and agree upon important details that otherwise can be just brushed over in verbal discussions. A written agreement should reflect the complete final structure of the business relationship between the people involved, leaving earlier ideas or solutions behind, creating a record of all the final details. A written agreement allows anyone or everyone involved to go back for a reminder whenever it is necessary. Everyone directly involved should sign and receive a copy of the final agreement.

It becomes more important to have a written agreement to fall back upon once a high-grade gold deposit is being developed. Because that is often when details become more important than they were before you found high-grade.

***"Lets not worry about how we're going to split up the gold; we'll figure that out if and when we find any..."*** This might work alright on other things, but with gold it is very poor organization. Arrangements like this can very easily end up in dispute once high-grade has been discovered. Believe me: ***No one is looking at the world quite in the same way after a rich gold deposit is added to the picture!***

Additionally, any agreements that you make in a gold mining venture should be tailored to suit the valuable amounts of gold that you are going after. The reason for this is that if you make a deal which seems to be fair with everyone based upon dredging a marginal pay-streak, and you then discover a very rich pay-streak that is producing commercial volumes of gold everyday, the deal may no longer be fair to everyone involved. But to try and raise the inequity with your partners and to get everyone's agreement on a new deal is almost impossible once large quantities of gold have been brought into play.

I learned this the hard way with an experience of my own. I had been dredging in an area for quite some time and had really put out a lot of energy to get familiar with it. Through dredging many sample holes, I was steadily working my way closer to a very large and high-grade pay-streak. But I did not know it at the time, because I had not found it yet. Man, I had gone through some real lean times prior to that. It was winter dredging with storm flows (fast, cold water), and I was trying to learn how to dredge and how to find pay-streaks – all at the same time, while living in a tent out on my mining claim. And I was just starting to get pretty good at sampling. But I didn't know that either. That is when an old friend of mine asked to come out and dredge with me for a few weeks. Well, he was a good friend, and his company sounded pretty good, so I encouraged him to come.

We worked out a 48/42% deal, with the larger share going to me. This seemed about right; because at the time my friend arrived, I was not finding paying-quantities of gold, and it seemed like he ought to have a fair-sized split of the smaller amounts of gold that we found together. We agreed that I should receive a larger share, because it was my operation and he was just visiting.

We hit that rich pay-streak on the first sample hole of the first day that he arrived! It was not the first pay-streak that I had been into, but it certainly was by far the largest and best-paying - even after our split.

But the thing about it is that I was planning to put that sample hole there as my next project anyway. I had been working right into that pay-streak for months in my data collecting, hard work and sampling activities (that pay-streak and half-dozen others for that matter, which I didn't know then, either). Man, were we into the gold; it was incredible!

I had invested the better part of a year's work into learning how to dredge and find that first big deposit. That was the gold which should have paid me off for all the hard times I had stuck through up until that point. Because of the deal I made with my friend, I ended up giving almost half of the gold to someone who took no part in the hard work of finding it. Sure, he helped clean it up. But cleaning up a rich pay-streak is the fun and easy part. The amount of time and energy it takes to get familiar with an area and learn how to find pay-streaks is worth a lot more. But this is something that my buddy would never have been able to fully appreciate, because he never had to go through it. To him it was just good fortune. For me to have tried to explain differently would have caused hard feelings. He would have figured that I was just trying to squeeze out of the agreement we made together. A deal is a deal. So he received his full split from the entire pay-streak and went away very happy. Me? I learned this lesson well: ***Make your partnership deals based upon the high-grade amounts of gold that you are going after and the correct exchange that should go to each person when that much gold is being recovered!***

I personally try to avoid forming full partnerships where the others involved receive some portion of the net profits – meaning the money that is left over after all expenses are paid. There are several reasons for this: (1) Putting yourself in a situation whereby you have to account for and gain the approval of others for every financial expenditure can really slow things down. This situation also creates the potential for internal disagreement and conflict as even the simplest of things can get micro-managed by a committee. (2) I do not like to sell my share of the gold unless I have to. We have gotten around this in full partnerships by splitting off some portion of the gold-recovery amongst the partners and only selling the amount of gold needed to pay expenses.

If you are going to do a full partnership arrangement whereby you will find yourself having to account (to others) for every expense you make, I suggest you keep the number of people you must account to as small as possible. The more opinions, the more potential for conflict.

It is easier if you create limited partnerships whereby you just pay a fixed royalty percentage of the gross recovery to those who are helping you. This way may is a bit more risky, because fixed royalties do not allow you to pay your expenses first. But at least you will not have to devote a bunch of your creative energies accounting for and justifying to others about every move you make.

You will have to make your own choices about how to structure your deals with partners. My advice? Keep it as simple as possible!

It is easier, by the way, to make a generalized deal that you are not particularly happy about, or in which all the details are not fully worked out, before you get into the gold. This is actually quite common. A few friends get together to go out and have a little adventure. They are not comfortable about discussing the gritty details of a business deal together, because they do not want to make anyone else mad at them. This is easy to do if you are not sure that you are going to find gold in paying amounts, anyway.

Then, after the gold is discovered, you really wish that you spent more time doing it right in the first place! I've seen this happen time and time again. It is <u>so</u> easy to talk glibly with your partners about how you will not have <u>any</u> trouble dealing with the gold, when *"the gold"* is still an abstract concept that you are doubtful will ever become a reality.

It is a whole different world when you are pulling pounds and pounds of beautiful, raw gold out of the river. You feel entitled to so much because of your contribution; your partners feel the same way because of their investments; and it all adds up to more than 100% of the gold! This is when the trouble really starts within the partnership. But it also is the time when the partnership should be turning up the production!

You can lose good friends over the smallest of disagreements when it comes to gold. I have watched it happen time and time again. It is better to work it all out right at the beginning; or if you cannot agree on the terms, just don't do the project together.

For that matter, when dealing with substantial amounts of gold, it is tough enough to prevent dissention within a group even when the best and most carefully-structured deals are made with the most ethical of partners! This is just the way it is with gold.

A number of things need to be discussed when structuring how two or more people will come together in a mining venture: *What part will each person play in the partnership? Who is going to be the project manager? How will you acquire the equipment needed for the operation? Where is the money going to come from to pay operating expenses and repairs? Who is going to handle the money? Exactly how and when is the gold going to be split up once you get into it? How are the nuggets going to be split up? Who, if anyone else, is going to be allowed to help develop any high-grade that is found? Who, if anyone else, is even going to be allowed to know about the pay-streak before it is cleaned up? How much time is going to be put into production? How much gold will be sold to handle expenses and repairs? Where the gold is to be stored before it is split up? How long will the partnership last?*

Sometimes agreements need to be modified to cope with unanticipated circumstances or some other situation that requires an adjustment. There is nothing wrong with modifying an agreement as long as everyone involved goes along with it. When this happens, the written agreement should also be modified as soon as possible, with everyone signing the modified version. It is good administration of business to make sure the written agreement continues to accurately reflect the existing arrangement between the partners.

## SPLITTING UP GOLD

My various partners and I have worked out a method of splitting up gold over the years which has always seemed to work out very well. On group ventures where everyone involved is getting some share of the gold we recover together, all recovered gold is deposited into a group collection jar. We pull out any and all pieces (nuggets) of gold that are of extra value, and put those into a separate group jar. We split up the nuggets at the end of the venture, because there will be more of them then, so everyone is more likely to get a more fair portion of them. The rest of the gold (the fines and flakes) are split up on a regular basis. We do it every day if there is enough gold to make the effort worthwhile. Sometimes we classify the gold through window screen, and measure up shares by weight of each classification to each partner. If any partner is to receive some lower or greater percentage of the recovery, the person will receive it in both classifications. This is good, because each partner gets something back every day from the work he or she has invested.

At the end of the venture, when it is time to split up the nuggets and jewelry gold, we pick out the most extraordinary nuggets and set them aside for the moment. Then we make equal piles of all the remaining pieces of gold as equal by weight as possible, and equal by size and apparent-value of the nuggets involved. Sometimes this can be a bit challenging, because each nugget is different. But when everyone is satisfied that the piles are all as equal in value that we are going to get them, we draw cards or tickets. Whoever has the highest card or first ticket gets to choose the first pile. The second-highest card gets to choose the second pile, and so on.

In our program, if one partner has some slightly-higher percentage of gold coming to him, he will usually receive the proper amount out of fines and flakes which were set aside for this purpose.

Then we take the most extraordinary nuggets that were set aside for last, and bid them off one by one - just like in an auction. Everyone involved in the partnership has an equal opportunity to bid up the amount of fine and flake-sized gold they are willing to pay for each nugget. Each nugget goes to whoever is willing to pay the most. The gold payments are directed into the group jar. Once all the nuggets have been sold off this way, the final gold from the jar is split up properly amongst all the partners according to our agreement.

While there are plenty of other ways to do it, this system has worked out very well, with no hard feelings amongst the partners, even though we often find very high-grade gold deposits with a lot of nuggets.

## DEALING WITH CLAIMS

Dealings with the owners of mining property should also be accomplished in detail before a dredger starts sampling around on those particular mining properties. This is for the gold dredger's protection. I have seen it happen quite a few times where a dredger makes a generalized verbal agreement with a claim owner, whereby the dredger is given permission to prospect around on the claim with the idea that they will make a *"fair deal"* if anything valuable is discovered. A deal like this might feel more comfortable than hammering out the details of a structured arrangement before gold is found. But things always look entirely different to everyone involved once a good pay-streak is located on the claim.

Before a pay-streak is found, the claim does not necessarily carry any great amount of value to the bargaining table, because no one knows for sure if there is even any pay-dirt on the property. This is the best bargaining position for the dredger. Under these conditions, a ten percent deal is very standard within the industry. That is 90% to the dredger. What I mean by this is that a royalty of 10% of the gross gold-recovery will be directed to the owner of the property that is being mined.

Once a rich pay-streak is found, if a specific percentage deal has not already been made, the mining property suddenly becomes very valuable, and the claim owner is in a position to be asking for just about any percentage that he chooses. His position will now be that a treasure has been located on his claim, and all that is needed is for someone to develop the deposit. And, of course, he is right!

I have seen it happen that a general deal was made between dredgers and a property owner *"to be worked out in detail if pay-dirt is found."* Once the dredgers located the pay-streak, the claim owner believed he was entitled to 50% of the gross gold-recovery. *"After all, it is rich pay-dirt on my claim, and all you have to do is dredge it up: 50% - take it or leave it!"* A property owner has the upper legal hand until you make some kind of firm agreement that allows you the right to develop the gold deposit(s) which you discover on his property. So it is not a good idea to strengthen a property owner's bargaining position by finding a pay-streak on his claim before you finalize a percentage deal. Some final arrangement should be made <u>before</u> sampling is started, with the value of the mining property taken at face value at that time.

Just like in partnership deals, the arrangement with a property owner should be thoroughly discussed: *Is the royalty going to be ten percent of recovered gold weight? Or, is it going to be ten percent of the complete range of nuggets, flakes, and fines? Is the split to be done on a daily basis? A weekly basis? Monthly? At the end of the season? Does the owner want to see the gold split up? How many dredges can you use on the property? Does the agreement allow you an exclusive right? Or might you find yourself competing with other dredgers on the same property?* These are the types of things that need to be agreed upon during the very early stages.

The time-frame is another thing that should always be discussed with a property owner. You should be comfortable that the property owner is not going to suddenly decide that your time is up on the claim, until after you have had a reasonable opportunity to mine out the high-grade deposit(s) that you have discovered there.

One of the richest gold deposits I located in my early years was turned over to one of my competitors by the claim owner, because my competitor made an offer to give the owner 20% of the gold-recovery (from a deposit that I discovered). My deal with the claim owner was 10%, because no underwater deposits had ever been located there before my arrival. But because there was no discussion about my exclusive rights to what I discovered, or how long I could mine on the claim, when my competitor offered the owner a higher percentage, I was told I had to stop mining immediately and move on. That was real disappointing!

If you are not receiving an exclusive right to dredge the property, it is very important to work out an arrangement that at least allows you an exclusive right to develop any pay-streak that you are able to locate on the claim. Build some protection into the deal so you can be sure that a competitor is not going to be allowed to move in on your pay-streaks, or that no one else is to be allowed to move onto the claim up in front of you and cloud you out with their tailings.

Competitors come crawling out of the woodwork once a rich gold deposit has been exposed. Expect that anyone who can figure out how to get some part of it will be in there trying their hardest. The main idea behind structuring good agreements in advance is to minimize the amount of confusion which results when high-grade is brought into the picture. Well-structured deals with your partners and the property owner will usually allow you to get directly to the business of recovering gold.

It is not always necessary or possible to get agreements in writing with property owners. Some people simply do not like to commit themselves to anything in writing, and you cannot push it without causing the person to tell you to just get lost. This does not necessarily mean that these people are lacking in integrity. There are some people around who may consider that there is probably something wrong with your integrity if a handshake is not good enough to seal a deal. You will have to find your own way through all this as you go along. However, the more you put at stake on a single mining property, the more important it is to have a written contract.

These are the rules of the game that you will be playing. So it is wise to discuss them long enough, and thoroughly enough, to insure that everyone directly involved understands all the details and are in full agreement with them.

## PERCENTAGE DEALS

In suction dredging, for the most part, at least in most of the places where I have operated for the past

25 years, the property owner usually receives ten percent of what a dredging operation recovers off of his property. I am talking about 10% off the top of the gross recovery.

I personally feel that it is excessive to pay more than 10% to a property owner if I have to locate the pay-dirt myself. If the property owner can tell me right where the high-grade has already been located, and it is just a matter of my going in there and developing an established gold deposit, depending upon how rich the location is, maybe a higher percentage might be justified

This mostly comes back to basic economics. You are paying for the mining and support equipment in the first place. You are going to pay the operating expenses, the repairs on your gear and the living expenses for yourself and your crew. You are doing all the hard work. You are taking all the risk. This is all on the chance that there might not even be any high-grade on the property in the first place. Twenty percent is too much of a demand by a property owner under these conditions. But some will ask.

One thing to keep strongly in mind while negotiating a claim deal, is that the bigger piece of the pie you agree to pay the property owner, the less percentage that you will end up with as your own reward after all the other expenses are paid. ***This means that the more you agree to pay, the richer the gold deposit has to be in order for you to make anything for your effort.*** While giving away a higher percentage might not sound like a big deal before you sample an area, just remember that there is <u>a lot</u> of ground around that will meet your minimum requirements on a 10% deal, which might not meet those very same requirements when you have to pay 20% to the property owner.

**Please keep this in mind:** We are talking about giving away a portion of the total gold-recovery. You will still have to pay yourself back for your investment in the mining gear, set up of the operation, repairs, operation expenses, and then split what is left over with the other people who are helping you on the mining operation. In a marginal pay-streak, the only person who might be making anything from all your hard work is the property owner! So it is a big deal to agree to give away some higher percentage of your gross gold-recovery to the property owner or anyone else. ***When someone asks for an additional 5% or 10% of your gross recovery, the number does not sound like very much; but in the end, that person could be making a lot more gold from your mining operation than you are!***

A few years ago, a person hired me to help him get set up a commercial dredging operation in Cambodia. The guy was very excited about his prospects for success, based upon some very positive results that I obtained in different area of the same country. But in order to gain all of the necessary permits and licenses over there in a hurry, he ultimately agreed to give away 90% of his gross gold and gemstone recovery to various officials in the country. And he had not even sampled for a deposit yet! I didn't find out about this until I got over there and got into a discussion with him about what percentage of gold he was planning to give the guys who would operate the equipment. He was asking me to help him arrange an experienced dredging team from the west. His idea was to allow the crew to split 2% of the gross recovery. Here is an example of someone who liberally gave away the whole pie! Nice guy! Because he had already made his deal with the officials, they were unapproachable on the subject of a renegotiation. Since he could never get help from an experienced crew on this type of deal, the operation never even got started. But the guy did end up losing all his gear and his full life's savings trying to set it all up. Moral of the story? ***The more of the pie (especially the gross recovery-pie) that you give away to others, the less chance your operation will succeed, and the less chance the other pie-holders will get much out of it, either.*** For a claim owner to ask, without a good and visible reason, for some unusually-high percentage, he is inviting the dredgers to high-grade on him and he most likely will end up receiving less than what would be reasonable. I have talked with a lot of miners over the years, and have found that for the most part, unreasonable property owners usually get treated unreasonably!

## KEEPING QUIET

One good way to get yourself into all kinds of trouble, is to start yelling and screaming out your excitement for the whole world to hear when you have just made some kind of gold discovery. This is just as true even it is just a little find - or even if you are just goofing off! Those people who hear you and your excitement imagine gold finds and bonanzas probably far better than what you are actually into. That includes the property owner. If you make big noise about a small find, word can get around, and the property owner might start wondering if he is being cheated.

On the other hand, if you find something nice and keep kind of quiet about it, it will be a pleasant surprise to the claim owner when you pay your royalty and he gets more than usual.

Please do not misunderstand me. I am not saying to withhold information from the property owner. I am advising you to not allow your own excitement and enthusiasm to create false perceptions which can thereafter get you into trouble.

When a person suddenly starts to do really good at something, there almost always seems to be someone else around who wants to make some trouble. This can be especially true when you start turning up high-grade gold deposits! It is better to not advertise for trouble!

When you are into a high-grade pay-streak, as hard as it may be to do, it is good policy to keep pretty quiet about it until after you have finished cleaning it up. Then you can make all the noise you want to. But while you are into it, if you do not tell anyone who doesn't need to know, you are less likely to become distracted by others who would like to get in on a piece of your action.

## INTEGRITY AND GOLD

Integrity is a personal matter. We each make our own decisions about how we will relate to others and the world around us. There are not many true saints around. Most of us have limits, whereupon weaknesses or flaws in our character allow us to fall from grace (when we do not do things as we know we should).

As most of us also are not monsters, we each also place personal limits to which we will not step beyond when it comes to the bad stuff. For example, a person might bend the truth around a bit, but would never steal something that belonged to someone else. Another person might steal small things, but never anything of great value and never from anyone the person knows. Another person might be willing to pull off any kind of scam to cheat others out of money, but the person would never kill anyone. Perhaps someone in the business of murder would draw the line when it comes to killing someone in his gang or family. We each set our own upper and lower limits by internal decisions. Life-situations push us up against these limits and we are tested.

***There are few things in life that will test your limits more than pure, beautiful gold.*** So it is wise in the beginning to do some introspection and decide where your personal limits are going to be.

***Anyone who believes they are not going to be personally tested by gold just has not found enough of it yet!***

I am not a saint, either. So I instinctively knew when I was just starting in gold prospecting that I had some internal decisions to make. Since I really enjoy the activity, and wanted to carve out a place for myself in this industry, I made my decisions early on that this was one area in my life that I was not going to mess up for myself through bad business dealings or unethical behavior. At least in this one area, I decided that I was going to try and do everything the right way.

Therefore, I personally never agree to a percentage deal with partners or property owners that I do not fully intend to pay. As part of this, during the time it is in my possession, I never allow myself to get attached to that portion of the gold-recovery which belongs to others.

Over the years, I have run across people on a regular basis that love gold so much, that I just cannot trust them with it. ***Gold is great. But you cannot let it take you completely over.*** In the end, it is a personal decision how deep you are going to allow yourself to get sucked in by gold. Moderation is the key to this.

Any game or activity in existence must follow an agreed-upon set of rules, for which there is a certain amount of freedom to act within. These are the same rules that are being formed up when a business agreement is made with partners or the owner of a mining property.

You cannot really win a game through cheating. To knowingly break the rules of a game so that you can win, is to admit that the game - by the rules - is above your ability to play. A person has actually already lost the game at the point where he decides to cheat or change the rules without the agreement of the others who are playing.

When a person finds him or herself making all sorts of justifications to reinforce some unethical behavior, the right and wrongness of the action must already be in question in that person's own estimation. It is the recognition of questionable behavior that prompts all the justification.

This discussion is not about how other people feel about you. It is about how you see yourself. I am talking about the internal mechanisms that people use to bury their own ability to see the truth in things.

Natural ability and intelligence fundamentally comes down to being able to recognize the differences between things, and to be able to act upon the differences. Whatever that quality is inside of you that can tell the difference between black, gray and white, or the differences between a hundred different shades of gray, cannot be separated from that part of you that is able to see the differences between right and wrong, or the many different shades of right and wrong. We are talking about the same thing.

A floor cleaner can tell the difference between a dirty floor and a clean floor. If he is good, he can tell the difference between a clean floor and a sparkling floor. And by being able to differentiate in this way, he will be able to learn exactly how to keep the floor sparkling. That is what makes him a good floor cleaner. This ability stems directly from the ability to see differences and the willingness to act upon them.

A gold dredger must be able to tell the difference between hard-packed material, and slightly-less hard-packed material. You need to be able to recognize when different layers change. You need to be able to see even the slightest differences in the amount of gold that is present. Prospecting is entirely about recognizing and comparing the differences that you see as samples are dredged into the streambed. Recognizing the differences, and then acting upon them, can mean the difference of finding a rich pay-streak or missing it entirely even when all the signs of its presence are right there in front of you. There are hundreds of tiny signs and things to see and know to succeed well in this field. The good gold dredger sees many of these little differences and is able to follow up on enough of the right ones to strike pay-dirt often enough to make the activity worthwhile. This is ability that I am talking about, and it depends entirely upon your capacity to tell the differences between things and act upon them.

Personal integrity and intelligence go hand in hand with each other. When you shut down your ability or willingness to recognize the finer shades of right and wrong, you also close off that same part of your ability to see important distinctions that allow you to follow the path of gold into high-grade deposits. Your search for gold through prospecting is really a search for the truth of the way things are.

No matter how you try to make yourself feel better about it (justification), when you cheat a claim owner or partner out of that portion of the gold that you have agreed to give him (them), you are burying that part of yourself that can see the truth of things. You are making yourself more stupid.

So the best answer in gold prospecting is to never cheat. The answer is to be very careful to only make deals that you can honor, and find enough gold that you can more than meet your obligations to the others who are involved. People who are good at gold dredging do not need to high-grade to succeed and do well at it.

Kick these concepts around for yourself and figure out where you are going draw your own line in the sand. Because if there is anything in this world that will put your personal integrity to the real test, it is gold! Gold can be very difficult to let go of once it is in your hands. Gold is especially hard to give away when, internally, you do not feel that the recipient really deserves to have what it is coming to him - even if you have agreed to it beforehand. In such a case, with gold, sometimes it might be tempting to not follow through on an earlier agreement. You can find yourself coming up with all kinds of reasons (justifications) as to why you do not need to follow through with the earlier agreements. Justifications can be made to sound very reasonable (to yourself). But if they are that reasonable, why not take them up with the claim owner? ***Gold is a very tempting substance. You need to be careful to not lose yourself because of it!***

I always try and work out a complete deal before I go onto a claim. There is much to lose by not doing so. If I cannot come to acceptable terms with a claim owner, it is much better to know ahead of time. There is no scarcity of rich gold-bearing river property around. I stay away from bad deals. This way, I do not put my personal integrity at risk in a gold mining venture.

I also stay away from making deals with bad people. Such people have a severely contagious disease. Take Joe Blow, for example, who is doing pretty well in life and making progress towards his goals. Then one day he takes up a partnership with Pete Schmuck, who is obviously a shady character. Let's say that Joe has taken a pretty strong stand on his morals, so he does not get pulled down into unethical activities by Pete Schmuck. One day, however, Schmuck finally ends up pulling the big con job on Joe Blow - or on Joe's associates, which is even worse! Joe gets mad and now feels the need to get even, or to stop Schmuck; or if nothing else, to make things right again with his associates - if that is even possible. But before Joe teamed up with Schmuck, he was doing fine and getting along well in life. See what I mean? Dishonest people are sick. They are losers. They are dishonest because they are not capable enough to do things the right way.

There is s saying, and it's a good one: "***If you want to fly like an eagle, quit playing with turkeys!***"

## GROUP DREDGING

After having observed many different kinds and sizes of dredging operations, and in having taken part in different-sized group operations myself, I have pretty-much come to the conclusion that gold dredging, for the most part, is better done as an individual activity, or as a very small group activity - like two persons, rather than as a larger group activity. There are exceptions to this, like when a group comes together for educational purposes or to share a common adventure.

Generally speaking, though, if you know how to find and recover your own high-grade deposits, and if you already have your own equipment, you will usually end up with more gold of your own in the long run if you do it on your own or with a single partner.

As a good prospector, you should be able to keep yourself into gold rather continuously. A pay-streak on any fairly large-sized waterway (river) is likely to run 15 ounces, 40 ounces, 80 ounces, or even more. I have seen many pay-streaks that produced hundreds of ounces. It may take a dredger two weeks to clean it up, four weeks, or maybe even the entire summer or several seasons. When you get yourself into a pay-streak, you are going to be busy recovering good gold for awhile; and what you recover will mostly belong to you.

When an individual is trying to keep several dredges into high-grade gold while working with multiple partners, he is only going to receive a fraction of the total recovery, and the pay-streak is going to get mined up several times as fast. If some of the other dredgers are inexperienced, it is almost certain that some of the pay-streak is going to be missed and left behind under tailings. This means that the person must find more than several times as much high-grade to make the same amount of gold for himself in the long run.

The most challenging part (and most valuable) of gold dredging is in finding the pay-streaks. While it does involve work, it is usually not very difficult to dig up a treasure once it has been located.

An individual by himself or with a single partner may only need to find one or two pay-streaks during an entire season to keep himself (themselves) busy dredging up high-grade and accumulating a substantial amount of gold. If the same individual or 2-person team needs to keep a larger group-program into gold, he might have to locate five or more pay-streaks to accumulate the same amount of gold for himself. That might be enough to keep him busy alone for the next five years or so!

## PARTNERSHIPS

There are several good reasons to team up with one other person in a gold dredging operation. One good reason is safety. A dredging accident is less likely to be fatal if there are two people on the scene keeping an eye on each other. A buddy is also able to offer a huge amount of assistance when it is time to move gear around during the sampling stages. Still another good reason to have a partner is the moral support that two persons can give each other when you are just learning to sample.

It is true that two dredgers are likely to finish off a pay-streak around twice as fast, with only half as much gold going to each person. But it is also true that two motivated people tend to spur each other on to get more accomplished - as long as both of you are hard workers. It can be much easier to locate the pay-streaks when two people are sampling and seeing it through together. And one other thing: In gold dredging, there can be some amazingly-good moments which carry an incredible emotional impact - like when a very rich deposit has just been discovered and the first bit of high-grade is being dredged up the suction nozzle. ***These are experiences which could never be wholly communicated to and understood by most others, so it is nice to experience them alongside someone else.***

Most pay-streaks can be dredged up with two dredges working side by side. To try and mine pay-streaks with more than two dredges at once can become unwieldy. It can be difficult to mange a third dredge (knocking out plug-ups or whatever) when it has other dredges tied to both sides of it. Under the majority of conditions, two-person teams are better than larger-sized group dredging programs. By that, I mean that the individual dredgers are likely to end up recovering more gold.

## PARTNERSHIP DEALS

There are several ways that two people can team up in dredging: They can form a full partnership and find, recover and split up all the gold equally between themselves. They can form a more limited partnership, helping each other to locate the deposits; and then clean them up together, with each person keeping what he or she dredges up. Or, they can form an even more limited association whereby they agree to help each other move gear around when needed, with each person finding and recovering his own deposits. Each of these different arrangements has its own advantages and disadvantages, and each can be most-optimally used under different sets of circumstances. It is good to be able to shift from one of these arrangements to another, depending upon what the situation is.

Teaming up with one other person on a full 50/50% partnership basis is done best when both dredgers are of relatively equal dredging and sampling ability, when both are willing to put equal amounts of time and energy into the operation and when both have about an equal share of financial investment into the program. Just by nature of the way gold affects people, if the gold being produced by an operation is going to be equally split up, then the amounts of input into the operation should be reasonably equal, too. Otherwise, dissension can occur between the partners.

The more gold that is being recovered, the more that the inequities tend to express themselves. This can happen in varying amounts, depending upon the relationships, how much gold is being recovered and how far the exchange of time and energy has gotten out of whack between the people involved.

I was dredging on a 50/50% deal with another dredger in a moderately rich pay-streak not too long ago. We each had our own dredges in the pay-streak side by side, but I was recovering two to three times more gold every day than he was. This can happen once in a while in a pay-streak when one guy will uncover a richer section of pay-dirt, but this was happening every single day!

The water was very cold, and it was towards the end of a long and hard-working, successful season. We were both kind of burnt-out, but I had set a quota for myself to put another pound of gold into my poke before knocking off for the season. The gold was there; I just needed to dredge it up. My buddy needed the gold too, but he was more burnt-out than I was. The only reason why he had not called it quits for the season, he said, was because I hadn't.

One day we sat down and I told him that I only wanted one more pound before ending off; and that I wanted to go after it on my own - keeping what I recovered, and allowing him to keep what he recovered. Since the pay-streak was just as rich on his side as it was on mine, and since he was an ethical man and close friend, he could see the fairness in what I wanted to do. So he agreed, and we started dredging side by side, each of us keeping what gold we recovered. Well, you will never guess: I started recovering a little more gold than I already was, and my buddy started recovering about three times as much gold as he was before! The pay-streak did not get any better over there. We just changed a working agreement. This is fruit for thought.

## GRUBSTAKING

There are various ways that different people can contribute to a dredging operation to keep the input relatively equal. One person may have the equipment and the experience, while another person might have the money to finance the operation. Full partnerships under these circumstances tend to work better when both parties are equally willing to put time and physical energy into the operation.

Those kinds of deals where one partner is going to put up the money and end up with half the gold, while the other person does all the work, generally do not work as well. Be careful about agreeing to a partnership on an equal basis with somebody who is not going to put in his equal share of dredging alongside of you. Since you can really only put in so many hours a day at dredging, there is still plenty of time left over to clean-up, cook, repair gear and do the other necessary things around camp. So there is no good reason to give an equal share of the recovery to someone else to do these things for you (unless it is your spouse or mother). If you do, you will be feeling the pain out of your own pocket.

Grubstake deals seem to work out better when the percentage to the silent financier is kept down to about ten percent or less of the gross find. Unless it means not going at all, you should try and avoid having to give any more royalties (especially of the gross recovery) away than you positively have to. As mentioned earlier, the larger cut of the pie you cut out for others, the more pay-dirt you will have to pass up that would otherwise be acceptable dredging.

Unless some very rich area has already been located and just needs to be dredged up, an absent financier and claim owner will likely make more gold in the long run if the amount of gold the dredger needs to pay out is kept down to a minimum reasonable percentage of the recovery. This way, the dredger is less likely to go broke.

I met up with a guy in Alaska who had agreed to pay a grubstaker 50% of his gross recovery, because the grubstaker had sent along enough groceries to last the dredger all season. Now let's see; that's 50% to the grubstaker, 10% to the claim owner, and the dredger still had to pay his own travel expenses, fuel and other operating expenses and repairs. How was he going to make any money for himself? He didn't, and neither did the grubstaker! And the worst of it is that he was into a very nice pay-streak, which he had to leave behind for someone else to dredge up.

In some partnerships, one person may receive a higher percentage of the gold-recovery because he has put more into the operation than the other. Regardless of the agreement, and how right it is, when split-up time comes, and one person's pile is larger than the other's pile, both people are likely to feel a little odd about it. This comes into play more when large quantities of gold are being recovered, making the difference in piles even more noticeable. This is the way it is with gold. Plan on it.

On that note, when forming partnerships, and the contributions are not all equal, it is common to allow 10% of the gross recovery to whoever owns the mining property. It is also common to allow 10% of the gross recovery to whoever owns the dredge. This is an exchange for putting up the capital for the main elements required to start a mining program. If someone is putting up the money to keep everything going, rather than a percentage of the gross recovery, if you can swing it, it might be better to just agree to pay that money back (perhaps with a little interest) right off the top of the gold-recovery. Then the guys who are doing all the work usually split up the balance.

It is not unusual for a commercial dredger, who already has all the gear, a place to mine and all the capital needed to keep the operation going; but he still needs a helper to dredge with him. Very seldom do dredge helpers get paid a wage or salary. They are almost always brought in for some percentage of the recovery. While deals vary, it is most common to see this kind of limited partnership come together whereby the helper (junior partner) receives 25 or 30% of the gross gold recovered – right off the top. That, by the way, is a big piece of the pie, considering that the mining property and mining gear should each be allocated 10% of the gross recovery. This only leaves around 50% of total production for the senior partner to pay operating expenses and repairs – and have something remaining for his own time and effort. Unless the operation is recovering a good deposit, sometimes the junior partner might be the only one making any money on this kind of deal. Still, you have to offer the person a big cut to make him interested, or you will find yourself operating a commercial dredge all by yourself!

Sometimes it works out better to set up a split on sliding scale, whereby the percentage that goes to a junior partner is higher when the amount of gold being recovered is less. This allows the junior partner to get an immediate return for his labor that is large enough to keep him interested. Then, when higher-grade pay-dirt is uncovered, the helper's percentage gets scaled back. Not so much that he does not benefit from the additional good fortune. But enough to allow the senior partner (who invested all the money and took most of the risk) to be rewarded for his larger investment into the program. By the way, this kind of arrangement should always be worked out in advance of finding the higher-grade pay-dirt!

## OVERMANNING AN OPERATION

One non-optimum way to set up a full partnership, is by teaming up on a small or intermediate-sized dredge with a second person who does not have an additional dredge to operate. One dredge can only process so much material in a single day. Two dredges of similar size, coordinated properly, ought to be able to double what one dredge can do by itself, as long as both operators are hard workers.

If you take on a partner using a small or intermediate-sized dredge, without also having a second dredge for him or her to operate, you will probably make less gold than you would if you just work alone.

There are exceptions to this, like when developing a high-grade gold deposit in deep streambed material where there are a lot of cobbles to be moved; enough to keep a helper productive full time without getting into your way. Another exception is when operating a large dredge, when it is a two-handed job to move the nozzle around. This creates the need for a helper to move the cobbles and boulders out of the way.

When two people are put to work on the same dredge, they need to be able to process at least twice as much volume of streambed material through the dredge and recover twice the amount of gold. Otherwise, the arrangement becomes inefficient, and the team might be better off operating two separate dredges, or take separate shifts on the same dredge.

In theory, you might figure that if one person can move a certain volume of material with a dredge, then two people working together on the same dredge ought to be able to move twice as much material. Sometimes this is true. If conditions are right for it, a well-orchestrated team underwater can process much more than double what either person could accomplish on his or her own. I have devoted a substantial discussion to underwater teamwork in ***Volume 2*** of ***Advanced Dredging Techniques***, so I won't go into that here.

If there is not enough work to keep a second underwater person productive on a dredge, without the person cutting into the nozzle-man's production and slowing him down, then it is not a good idea to put a second man there. What do I mean by cutting into the nozzle-man's production? When a second person runs out of productive activity to keep busy, to stay busy, the person can start moving cobbles that are not ready to be moved. In doing so, streambed silt will be released into the water. Then, if not immediately sucked up, water currents can cause that silt to spread around and cloud-out the hole. Then the nozzle-man must wait until the water clears up enough again that he can see what he is sucking into the nozzle. This is just one example of how a second person can actually slow down the production in a dredge hole.

As an example of an overmanned (and therefore underpaid) production dredging operation, I teamed up with two other guys one time in a sampling activity to locate a pay-streak. We found a very nice one which was sitting underneath about two feet of hard-packed material over top of some really rough and irregular bedrock. The pay-streak was about 45 feet wide; so there was no problem placing our two dredges side by side to clean up the gold.

We were each operating 5-inch dredges. Our partnership agreement was that after working together to locate some high-grade, we would develop the deposit side-by-side, keeping our clean-ups separate. I was running my own 5-incher and keeping all the gold that I found. The two of them teamed up on their 5-inch dredge and were splitting what they found equally between the two of them.

The streambed material was very hard-packed, and the water was moving pretty fast. A deep orange-colored silt in the material made it necessary to take apart the streambed in an orderly fashion, sucking the silt and material from the same spot that you were moving cobbles. Otherwise, that part of the hole would be entirely clouded-out by silt, and you would have to wait for a half-minute or more before it would clear up enough that you could see what you were doing again. This is standard operating procedure in most dredging situations, anyway.

The hole we were dredging was so wide, that if I clouded myself out, their side would still remain clear enough that they could continue on dredging, and vice versa.

Since we agreed that each would keep what we separately recovered, and agreed that each could spend as many hours at it as we wished, the game was to produce like mad to dredge up as much gold as possible before the other guys recovered it. The friendly competition between us really spurred us on. Those were some hard-working guys and they surely kept me moving all the time, me knowing that they were dredging up all that gold as fast as they were. They had teamed up on their dredge and were going at it hot and heavy on the other side of the hole. But at the end of each day, my tailings pile was at least twice the size of theirs, and I usually had twice as much gold, too. Sometimes more!

In wonderment at this, I slowed down enough one day to have a look at what they were doing. Sure enough, they just kept clouding each other out. In 2 1/2-feet of material, there just was not enough work to keep two guys effectively busy on one dredge. I then went over and watched myself dredge, to see what a second person could do to help me get material through the nozzle faster, and there wasn't much. I was pouring it through about as fast as it could be done most of the time.

The best bet for those guys would have been to split the day into two shifts; one starting at daybreak, and the other one ending at dark. In that way, each one of them would have ended up dredging about the same amount of gold that I dredged. As it was, they probably would have about doubled their gold intake if one of them did not dredge at all! This is something to keep well in mind when organizing a dredging operation.

## WATCH OUT FOR THE CON ARTISTS!

There are three common areas in gold dredging where you may run into people who are going to try and pull the wool over your eyes, while they steal the money and gold out of your pockets. These are: mining claim sales, used mining equipment sales, and when you are selling your gold.

**BUYING CLAIMS:** Let's take up buying a claim first. They sell for just about any amount, depending upon how much value is being attached to them. Notice I didn't say: *"How value they really contain!"* Mining claims can sell for $200, or they can sell for a million dollars!

The best way to determine the actual value that a mining claim holds for you, is to thoroughly sample it. If you are thinking about buying a mining claim that you intend to dredge, it is not good enough to take samples up on the banks with a gold pan. It really takes going out there with your equipment and dredging sample holes to obtain an accurate idea of how valuable the claim is going to be to you as a gold dredger. It also is not good enough to dredge just one or two sample holes and stop there if you find pay-dirt. If you are going to pay a lot of money for the claim, it is wise to dredge more holes to see how long and wide the pay-streak actually is. Sometimes you can choose just the right spot to dredge (or the seller tells you where to dredge) a sample hole and put down in beautiful pay-dirt, only to find out later that it was a single deposit, or a very small pay-streak! This can be very misleading. I have seen it happen a few times.

If a claim owner is asking a lot of money for a mining claim, based upon his assurances that it is *"very rich,"* then he should have no objections to you doing some testing to see for yourself. It would be unreasonable for a seller to expect you to rely entirely upon his assurances.

Watch out for the seller who plays it up real big and tries to get you so excited about all the gold you are going to recover, that you do not feel the need to do any testing! *"Oh man, this claim is so hot that it kills me to sell it—but I'm too old to work it and I need the money ..."* This is where the con-man is going to try and get you.

When dealing with mining claims, if it is worth a lot of money, then it is definitely worth a lot of sampling first!

Another way that you can get stung on a mining claim deal, is by buying one from someone who does not really own it, or by paying royalties to someone who has given you permission to dredge on the claim, but who doesn't really own it.

With some research, you can locate any active mining claim in the County Recorder's Office and the Bureau of Land Management records, and find out who really owns it and make sure the yearly paperwork has been kept up. At the same time, you can research to see if anyone else has filed a mining claim over the very same location. If two or more people have filed mining claims over the same ground, future conflicts are likely, so it may be wise to look for another mining claim.

Perhaps it is not necessary when you first make a percentage deal over a claim with a person who apparently owns it. But if you get into large amounts of gold, it could be worth the little time it takes to verify who the actual owner is.

**BUYING USED DREDGES:** Before you buy a used dredge, it is a good take it out onto the river and run it to make sure that it operates properly. This might not be positively necessary when it is a used standard store-bought type of dredge, but it is definitely a must when you are considering buying a home-built dredging rig which you have not operated yourself. There can be many bugs that have yet to be worked out of a home-built dredge which might not be apparent until you take the dredge out and run it, especially in pay-dirt if possible. This is really the only way to see what you are buying.

Here are a few things to check on used dredges: *Will the motor start? Open up the pump housing to make sure a faulty bearing did not allow the impeller to chew itself up, along with the insides of the pump. When you start the motor, check to see if the hookah-air compressor is pumping out air. Check to make sure the plastic pontoons are not brittle from sitting too long in the sun. Is the suction hole stiff and brittle? Is the power-jet worn out?*

When looking at used dredges, I don't worry too much about a little rust. That is something that a wire brush and few coats of paint can take care of. I look for the things are going to cost money to replace or repair: Motor, pump, air compressor, power-jet, hoses, pontoons and air-breathing regulator.

Make sure to get a bill of sale from the person you are buying the dredge from, and write down the person's drivers license and auto registration details. The reason is that dredges do get stolen off the river every once in a while. If you buy stolen goods in California, you must forfeit the goods back to the true owner, even if the guy you gave the money to is long gone!

**SELLING YOUR GOLD:** There are plenty of honorable gold buyers around who will give you a fair deal on your gold. There are also plenty of people around who would prefer to give you a lousy deal on your gold.

Your best bet is to have a number of different buyers available to sell your gold to. Make some phone calls when you are ready, and sell to whoever is willing to offer you the most.

Unless you are dealing with an established refinery, the best way to sell gold is usually by the wait of what you have. For example, selling an ounce of placer gold at 80% of the daily spot value. If gold is selling for $500 per ounce that day, on an 80% deal, you should receive $400 for your ounce of gold.

I avoid dealing with gold buyers who want to start subtracting the value of impurities and getting into complex or confusing calculations. Most gold buyers in gold country already know how pure your gold is. This is figured in when you agree to receive a percentage of the spot price for the weight of your gold.

Actually, in the active gold mining areas, if you ask around, you will find that there is usually some pretty stiff competition amongst the local gold buyers. It will not take long to find out what the going rate is (percentage of the daily spot price for the weight of your gold) for the gold that is being mined in the area. As long as you clean up your gold well, there is no reason you should not get whatever the going rate is.

# 5

# ADVANCED RECOVERY SYSTEMS

The smaller that a particle of gold is, the more difficult it will be to recover in a sluice box. It has been estimated that a piece of gold weighing 1/4-grain can be divided into at least 2000 separate, smaller particles, each which is recognizable as a single color. There is an abundance of gold in any gold-bearing river which is so fine in size that it is almost completely dissolved into a state of suspension in the river water itself. You are not likely to trap a substantial amount of gold that fine in size in a conventional sluice box, because the gold is so small and light that it will flow right along with the water.

I know a very experienced old-timer who has excellent fine gold recovery on his dredge. One time when he was dredging up a pay-streak, for an experiment, he had a buddy hold up a 2 1/2-foot square amalgamated plate just behind his dredge where the silt would pass over it while he pumped material out of the pay-streak. He says that it only took a short while for the plate to turn yellow from the ultra-fine gold which passed over and out of his sluice box. The gold was extremely fine, so fine that there was only a small trace of it showing in his sluice box concentrates.

Any sluice box will effectively recover gold down to a specific size of particle classification. We can call this the *"effective size range of gold recovery."* A sluice will start losing more particles of gold when they are smaller in size than the effective size range. The smaller the particles, the greater percentage of them that will be lost.

No gold dredger likes to be told he is losing any gold out of his sluice box, especially when other dredgers are telling him how much gold he is losing. So you run into some miners out in the field who *"know"* their own recovery systems are the best thing going. This is ridiculous really; because anyone who's been around ought to know by now that my dredge's recovery system is the best!!

Actually, the real test is how fine in size your recovery system will continue to recover a large percentage of the gold particles.

There is a cutoff point for the effective range of particle size recovery. If you are seeking to improve your recovery, it will be necessary to work on improving the system's efficiency in catching more of the smaller-sized particles.

A dredger can pan down a set of his concentrates, see some ultra-fines in his pan, and get the mistaken idea that his dredge is recovering effectively down to that particle size. Almost any concentrating-type recovery system will catch at least some of the ultra-fines. The point here is that you cannot see the fines and ultra-fines which were washed right through and out of the sluice box. You can see, however, that you are recovering gold abundantly down to a certain size range. This will give you some indication of what your effective range of recovery is. To see some small amount of ultra-fine-sized particles in your concentrates just proves that they are there in the streambed material. The key question is: How much is there?

The purity of gold particles, and also the texture and shape of their surface, have a lot to do with how small in particle and flake-size that your recovery system will be able to effectively recover. The less pure that a specific particle of gold is, the lower its specific gravity will be. So the slower the water flow will have to be through the sluice for that particle to drop out and become trapped. As an example, your recovery system might recover 90% of the 100-mesh gold particles when it has a purity of 88% gold. The same recovery system - unchanged in any way - might only recover 10% of the 100-mesh particles which are only 78% pure. So when you move from one location to the next, if the purity, surface texture and/or shape of the gold particles in the new area have changed, the efficiency of your sluice box in fine gold recovery might appear to change, as well.

## RIFFLES & CLASSIFICATION

In gold mining, when we talk about gravity-recovery systems (as opposed to chemical-recovery systems), we are basically talking about the creation of a suspended medium. A suspended medium is a condition of fluidity where materials are allowed to separate because of their relative weights.

Since gold is around six times heavier than the average of other materials which make up a streambed, if you pass raw streambed material through the right kind of suspended medium, you are able to drop the gold to the bottom and direct the lighter materials off the top as tailings.

As an example of this, if you dropped some gold onto the hard ground, even though the gold is much heavier than the material making up the ground, because the ground is somewhat hard and compacted, it would probably take a very long time for the gold to work its way through to the bottom of the material in that compacted state.

But if you dug up all that ground (and gold) and dumped it into a raging river during a major flood storm, the material would get all churned into a liquid slurry as it is washed downstream in the river, and the gold would very quickly work its way down through to the bottom of the slurry. This is because the slurry is in a liquefied state of suspension, where heavier particles can easily penetrate downward because gravity is pulling on them much harder than the lighter materials.

There are different ways to create suspended mediums so that gold can be recovered from streambed materials. Mechanical jigs create a pulsating medium within a chamber that keeps water and material in a state of fluid suspension. Oscillating sluices create a left and right movement (similar to panning) as material is washed over top by a mild flow of water. Shaker tables use a finely-tuned vibrating action, with a very mild flow of water, to separate material by weight. Because these systems use mechanics to help create a suspended medium, we refer to them as *"mechanical recovery systems."*

Most recovery systems on suction dredges use fixed riffles to trap gold out of lighter streambed materials as they are washed through a sluice box by a flow of water. Riffles are baffle-like obstructions, fixed in place along the bottom of the sluice box. They are designed and positioned so that there is a back-pressure created that sucks water and material behind the riffles from the flow over top. When the correct water flow is directed over a riffle, the back-pressure keeps the area just behind the riffle in a state of continuous fluid suspension. Therefore, the heaviest materials are able to concentrate there.

Gold recovery systems that use a flow of water over riffles are called *"fixed recovery systems."*

***Classification (sizing of material) is the key to fine gold recovery!*** This is true in both fixed and mechanical-type recovery systems. In general terms, this means that the finer (smaller) in size that you want to recover gold efficiently, the more closely the material must be sized, and the more finely-tuned the suspended medium must be to facilitate the separation.

As an example to put this in perspective, the water flow it would take to move a 5-inch (diameter) rock would also likely be violent enough to wash away a fine particle of gold. And vise-versa: A suspended medium tuned so gentle as to allow fine particles of gold to efficiently separate from lighter streambed materials would be completely overwhelmed by a 5-inch rock.

The general idea is that if you want to efficiently recover finer-sized gold, you have to first separate it from larger-sized material. This is accomplished by screening. The process of screening is called *"classification."* Classified material(s) can then be directed into a recovery system(s) that has slower water and more gentle suspended medium(s).

What do I mean by a *"gentler suspended medium?"* In this discussion, I am mainly talking about the size of riffles. Because, the bigger (deeper) the riffle, the greater (and more violent) water flow is required to maintain a suspended medium of water and material behind the riffle.

Let's talk a little about what happens behind the riffle in a sluice box. If you only run water over the box, it does not take very much water flow to maintain fluid movement behind the riffle. This is because there is nothing solid there to obstruct the flow. But the average streambed material (rocks, gravel and sand) weighs around 4 times more than water. This means water flow and turbulence must be dramatically increased to keep material in a suspended state behind a riffle. The deeper the riffle, the greater the volume and weight of material which must remain suspended – so the greater the water force (and violence) needed to keep a riffle from packing up.

What is *"packing up?"* That is when you overwhelm a riffle with too much weight of material (rocks, sand, silt), and the suspended medium is lost. When the suspended medium is lost, the riffle will no longer concentrate the heaviest materials as they pass over the riffle.

Keep in mind that when a riffle is operating correctly, it will continue to concentrate the heaviest materials that are passed through the suspended medium.

At the beginning of a production-run, the specific gravity of the average material behind a riffle will be similar to the average specific gravity of the raw material found in the streambed. For the purposes of this discussion, let's say this is around 4 times the weight of water. But as other heavier materials like iron (specific gravity of around 8) flow into the riffles, the heavier materials will displace lighter materials, and the materials behind the riffle (called *"concentrates")* will become heavier. Each area is different, but it generally does not take long to accumulate concentrates in a sluice box that are around twice as heavy as the average raw material being processed. Heavier materials are usually associated with iron. They are dark-colored by nature. They are also often referred to as *"black sands."*

The heavier the concentrate behind a riffle, the more water flow and turbulence is required to keep it all in a suspended medium. The deeper the riffle, the more it will become overwhelmed by the increased accumulation of heavy material. So a deeper riffle requires a faster, more violent flow to keep it working as the concentrated material becomes heavier.

On the other hand, the shallower the riffle, the less force of water it takes to maintain a suspended medium behind the riffle. Less water force and violence will allow smaller particles of gold to settle.

But a shallow riffle also requires that the water flow be reduced. This is because a shallow riffle will not create enough protection behind it to prevent turbulent flows from boiling fine gold away. What do I mean by *"boiling?"* I mean that if you put too much water flow over top of any riffle, the suspended medium becomes so violent that even the heaviest particles can be washed away.

Here is what I am saying: ***If you want to recover finer gold more efficiently, then it is necessary to direct finer-sized materials into more-shallow riffles, that require milder water flows to keep them functioning when filled with concentrated material.***

But the problem with suction dredges is that we are sucking up a lot of rocks! Let's just use the example of a 5-inch dredge. The nozzle-restriction will allow a 4.5-inch rock to be sucked up into the sluice box. And there must be enough water flow through the sluice to keep the 4.5-inch rocks, and everything else that is being sucked up, flowing through and out as tailings. Otherwise, the material will build up on the dredge and sink it in very short order!

So initially, we are dealing with a force of water through the sluice box that will wash 4.5-inch rocks all the way through. That is a lot of force. And the force is turbulent.

Part of the reason why a rock is moved by a water flow is because it is being pushed along by the water. A nearly-equal reason is that it is also being pulled along by the vacuum that is created behind the rock as the water flows around it. The vacuum behind a rock creates enormous turbulence that will affect anything it comes in contact with – including small particles of gold. The turbulence associated with the movement of a 4.5-inch rock is a lot, compared to the mild state of suspension necessary to allow a fine-sized particle of gold to settle behind a short riffle.

Since we must suck up rocks in dredging (the bigger rock we can suck up, the fewer we have to move out of the way by hand), the first priority in the recovery system is to separate the big rocks away from where we want to recover the fine-sized gold. We do this by dropping the smaller-sized materials through a classification screen, and then directing them into slower-moving recovery systems. Sometimes, we are even able to screen two different-sized products, and direct each into a separate recovery system where the suspended mediums can be adjusted accordingly.

One common mistake I see with suction dredges, is that classified materials are sometimes being directed into slower-moving recovery systems that are using deep riffles. As these riffles do not have enough water flow directed over them to create the required suspended medium, they pack up and do not work very well.

Bigger is not better when it comes to riffles in a fine-gold recovery system. In fact, it is worse.

Here are some things to look for to get a better idea if a riffle is working for you:

1) The material behind a riffle should be visibly dancing during normal operation. You should be able to see that all of the material there is in a continuous state of vibration and movement. This can be deceiving, however. Because sometimes only the surface of the material behind a riffle is in movement. Everything under the surface can often be packed up.

This is almost certainly the case; if after you shut down, you scrape behind the riffle and find that just below the surface of black sand, it is all packed up with lighter, blond-colored material. In this case, you will have to decide if you need to increase the water flow and violence (which will make it harder to settle fine gold) or shorten the riffle. ***Shortening the riffle is usually the best solution.***

Please do not misunderstand me here. I am not talking about the riffles that get big rocks passed over top of them. I am talking about the riffles where you direct pre-sized, smaller material.

2) Do some controlled tests at production-speed (meaning while someone is sucking material into the dredge's suction nozzle), while feeding a pre-weighed amount of fine-gold into the system. You can either catch and test the tailings, or you can process the gold from the recovery system (or both), to evaluate how well your recovery system is working.

If you are losing gold from a fixed recovery system, it will come down to either a screening system that is not working very well (to separate the fine gold from the big rocks that must be washed away using heavy force), or a bad relationship between the flow of water and depth of riffles in the fine-gold recovery system. The answer is to just keep working at it until you get it right.

So, if bigger and deeper riffles are not the answer to recovering fine-gold from pre-sized material, how small and shallow should you go? Weight and shape-characteristics of the average streambed material in each area may be a little different. For example, crystalline, angular-type material requires more force and turbulence to keep it moving along. So there is not a fixed answer that will work perfectly in every single situation.

However, as a general guideline, I would say that the depth of a riffle should not be much more than the maximum size of material that is being directed there. In other words, if a 3/8-inch screen is being used to pre-size material, you probably do not want to use riffles much deeper than 3/8-inches. Then you set the water flow to keep those riffles fluid when there is an accumulation of concentrated material behind them.

**Remember:** If you use shorter riffles, you must slow down the flow over top of them to keep from boiling-out your fine gold. Slower water and less violence allows for finer particles of gold to settle – as long as a suspended medium is maintained behind the riffle. If you can keep the black sand in movement without boiling out the riffles, you will have a great recovery system. Because gold is around 2.5-times heavier than iron, it will displace the iron behind the riffles as long as there is movement there.

## LARGER-SIZED DREDGES

Classification becomes even more important as the size of the dredge is increased. This is because a larger dredge has the capacity to wash larger-sized rocks through the recovery system, which creates a greater amount of water force. Larger rocks will cause more water turbulence around, behind, and below them as they are being washed through a sluice box. This additional turbulence creates violence in the recovery system that can boil fine-sized gold out of the riffles.

A larger-sized dredge in experienced hands is able to process a greater volume of streambed material. If you are able to recover the fines and ultra-fines with a larger dredge, you can make lower-grade streambed material pay well enough to meet your minimum standards. If you can make lower-grade material pay off, there will be substantially more acceptable pay-dirt available for you to dredge in most of the gold-bearing rivers.

Classification in most dredge recovery systems today is being accomplished by washing the materials over top of a classification screen or punch plate (commonly called a *"classifier"),* and directing the classified materials out from underneath the screen and into a slower-moving recovery system. This is primarily being accomplished today in three different ways: triple-sluice boxes, double-sluice boxes, and screened-over riffles in a conventional sluice.

## TRIPLE-SLUICE RECOVERY SYSTEM

The triple-sluice has a classification screen at the head of the box, which allows smaller-sized materials to be separated out and directed into side sluice boxes where the flow of water is much slower. Most flakes and particles of gold are of a size which will drop through the classification screen and end up in the slower-moving side sluice boxes (See Figure 5-1).

Because the triple-sluice is a hydraulic-type of recovery system (water force is used to wash the larger-sized materials over the classification screen and through the sluice), some of the fines and ultra-fines will be carried along by the heavy force of water across the top of the classifier and washed into the center box. The center box will recover fine gold the same as a conventional single sluice box. Actually it might recover a little better, because not nearly as much volume of heavy concentrates will be washed into the center box. Therefore the riffle system in the center box will not load up as heavily with black sands as it would in a conventional single sluice box.

The main advantage to a triple-sluice is that a large percentage of fine gold will be washed into the side sluices where they will thereafter be subjected to a much slower and less violent flow of water. Therefore,

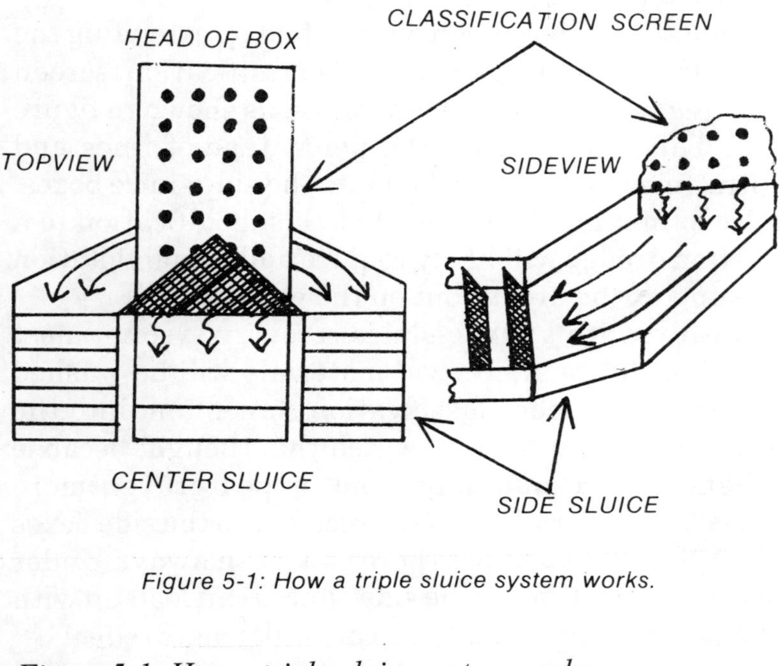


*Figure 5-1: How a triple-sluice system works*

the effective recovery of smaller-sized gold flakes and particles will be improved.

The best bet with a triple-sluice (and a double-sluice), is to try and direct as much of the smaller-sized materials down through the classification screen as possible. The purpose is to protect as much of the fine gold as you can from the violence associated with the main fast stream of water that is required to wash larger-sized rocks and materials through the recovery system. The length of the classification screen, before it reaches the side sluice box openings, can affect how much classified material is washed into the side boxes. The longer that the classification screen is, the more smaller-sized materials can be washed down through the screen and into the side boxes.

Triple-sluice boxes can be modified by extending the length of the sluice box and classification screen upstream of the side box openings, as shown in figure 5-2. This can increase the percentage of fines and ultra-fines that are washed into the side sluice boxes.

Keep in mind, however, that this modification to a suction dredge will likely require additional flotation to support the dredge out on the water.

Ideally, with a triple-sluice recovery system, most or all of the smaller-sized materials will be washed down through the classification screen and into the side boxes. You have to watch this though; because when you are pumping sand or pea gravel, more classified materials can be directed into the side boxes than the slow currents there can wash away. Under these circumstances, the side sluices can load up with too much material and you can sink your dredge!

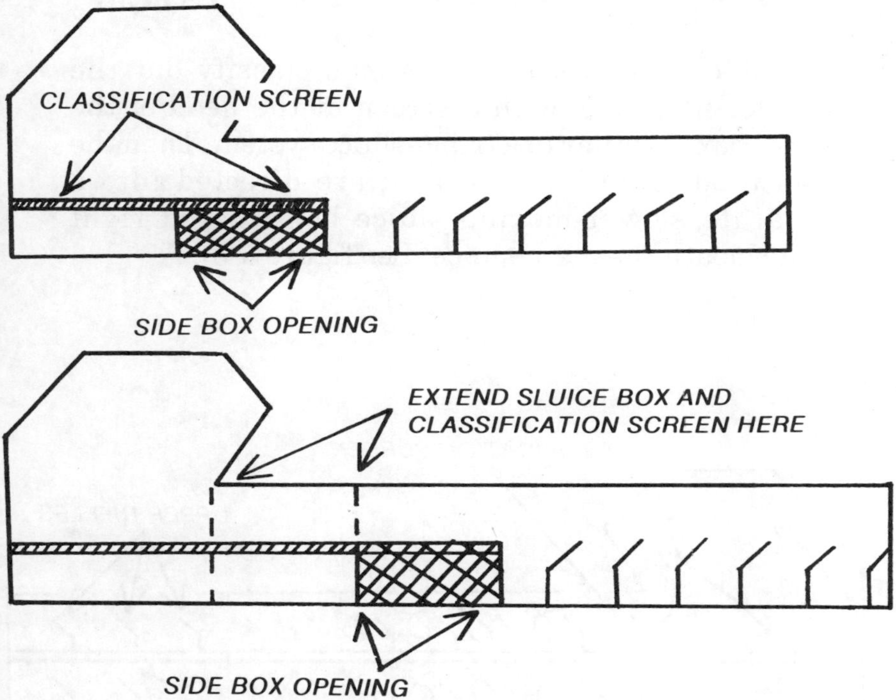


*Figure 5-2: Triple-sluice modification allows more fine gold to get into the side sluice boxes.*

When building or modifying a triple-sluice dredge, it is helpful to devise some sliding doors across the side box openings so that you can regulate the amount of water and material which flows into the side boxes - or shut them off altogether when you are pumping large volumes of valueless sand or pea gravel.

## DOUBLE-SLUICE RECOVERY SYSTEM

Double-sluice recovery systems classify out the smaller materials with a screen at the head of the sluice box – similar to a triple-sluice system. Then the classified materials and gold are directed into a separate slower-moving sluice box located right underneath the main sluice (See figure 5-3).

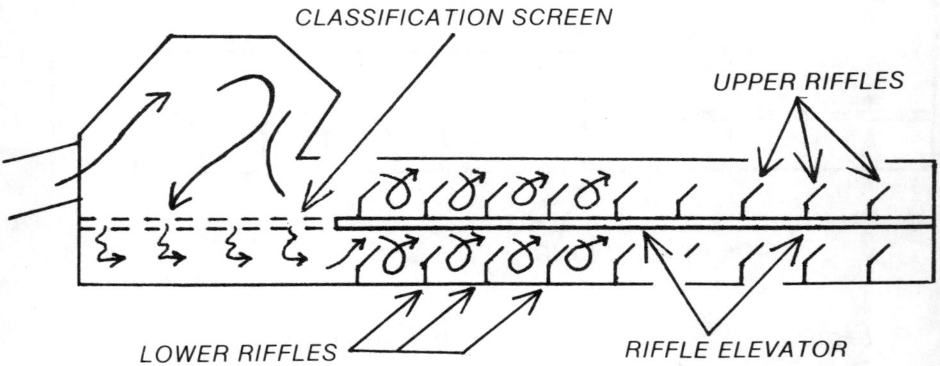


*Figure 5-3: A double-sluice directs classified materials and gold to a slower-moving sluice* <u>*under*</u> *the main sluice to improve gold recovery.*

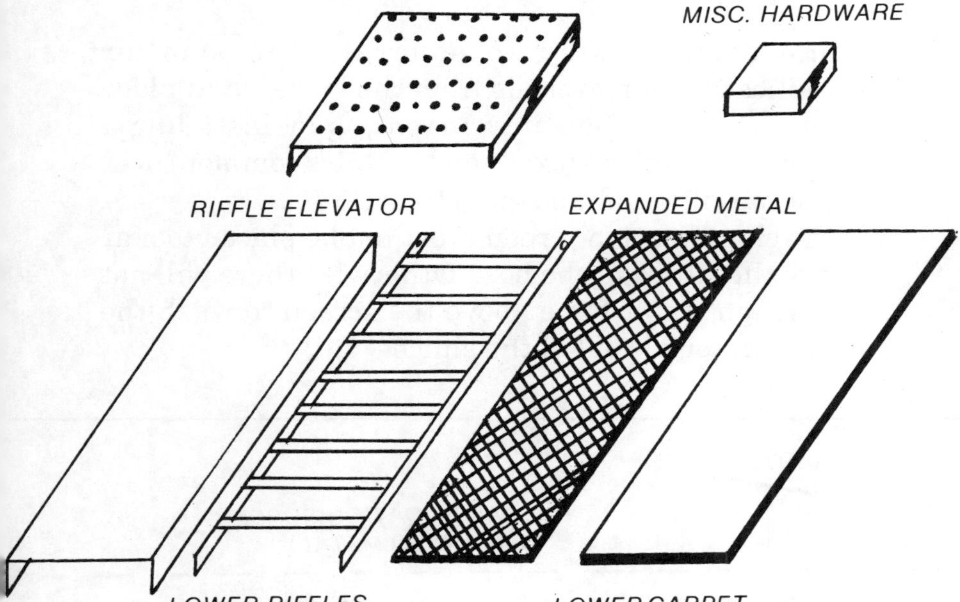


*Figure 5-4: Double-sluice conversion kit.*

Because of the improved recovery, some dredgers convert their conventional single sluice boxes into a double-sluice system. The components normally required to accomplish this are a new and different-sized classification screen for the head of the box; a new set of riffles with expanded-metal and carpet for the lower fine gold recovery system; a riffle elevator, which simply acts as a solid floor foundation that supports the upper standard set of riffles; and the miscellaneous hardware needed to install the system in a standard single box (See figure 5-4).

## SCREENED-OVER RIFFLES

One common way to separate the smaller classification of materials from the larger, and place them in a slower-moving current, is by installing a classification screen (expanded-metal is common) over top of the riffles, as shown in figure 5-5.

The classification screen must not be placed too far above the riffles in the box. Otherwise there will not be enough water force above the screen to wash the larger materials out of the sluice.

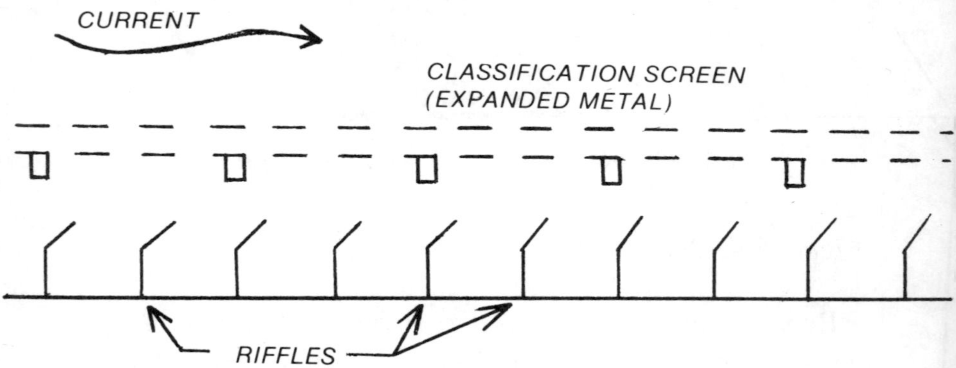


*Figure 5-5: A classification screen can be used to keep the larger-sized materials out of the riffles.*

The turned-up type of expanded-metal works well with this kind of system, with the turned-up points pointing <u>with</u> the current in the box, as shown in figure 5-6. They are less likely to jam with pebbles this way, and they also will not direct as much of the fast upper current down beneath the screen into the riffles.

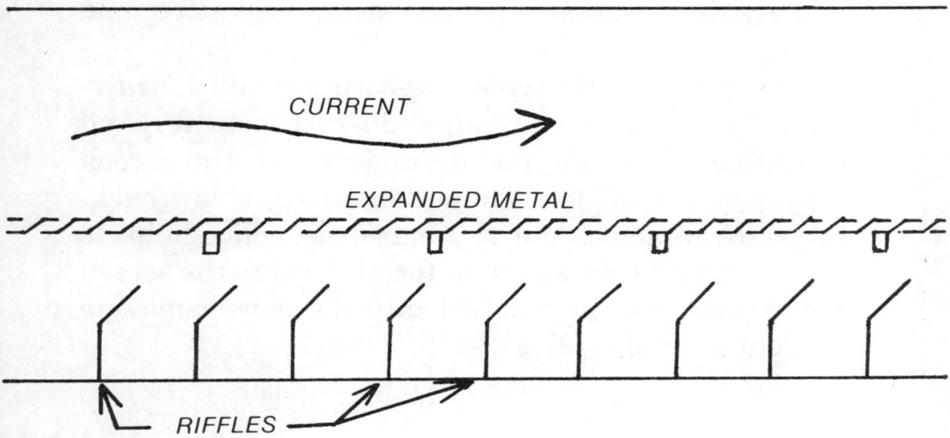


*Figure 5-6: When turned-up expanded-metal is used as a classification screen, the turned-up points should be pointing with the water's flow.*

To further slow down the water and material underneath the classification screen, cross supports can be extended just below the screen, and parallel to the underlying riffles. These cross supports *("water-dams")* should be a width of about 25% the distance that the screen is positioned above the riffles, as shown in figure 5-7.

How far above the riffles that the screen should be, depends upon the dredge size and the volume of water moving through the sluice box. This will vary from dredge to dredge. You will have to do some trial and error to find the balance whereby there remains enough water flow over top of the screen to wash big material out of the sluice, and enough water flow under the screen to keep the riffles from loading up – all while you are dredging hard-packed streambed at volume speed.

The reason why the screen supports are called *"water-dams"* is because they slow down the water and classified materials moving underneath the screen just over the riffles. They also create a low/back-pressure area just behind themselves, which helps to pull additional classified material through the screen where they can be washed into the slower-moving current over the riffles (See figure 5-7).

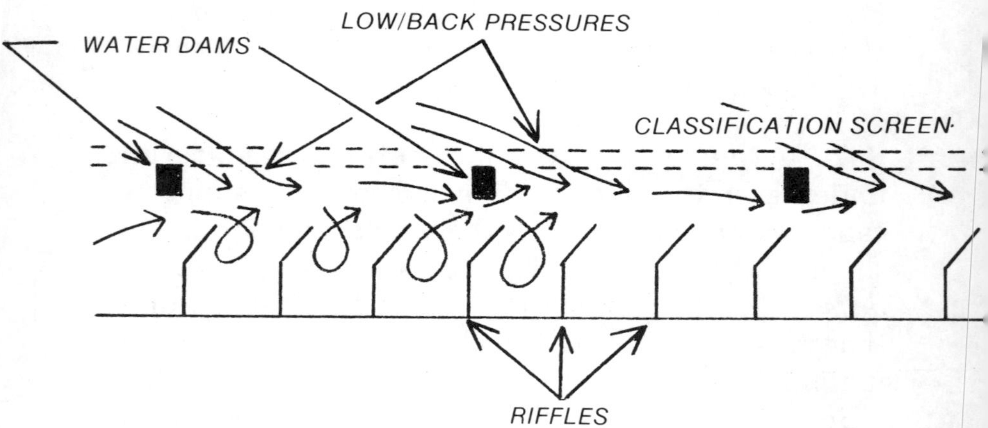


*Figure 5-7: Water dams slow down the current below the classification screen, set up a low/back-pressure area just behind themselves and help pull more material down below the screen.*

Note that it is important to leave some riffles in a sluice box open, so that there will be places for larger gold nuggets to drop out of the main flow of water.

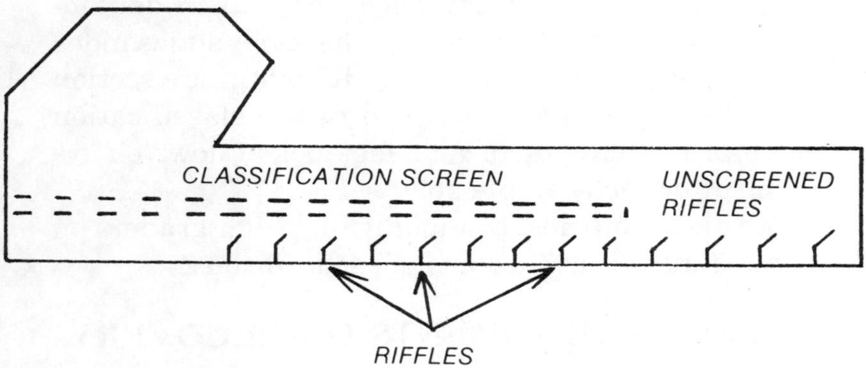


*Figure 5-8: Leave several riffles open to help trap the larger-sized nuggets, too.*

## HIGH-GRADE RIFFLES

It is common these days for dredgers to set up a separate smaller section of riffles at the head of their sluice box which can be removed and cleaned out quickly during sampling, or at the end of a production day. This special section of riffles is often referred to as the *"high-grader."*

The best place to set up a high-grader is right at the head of the box where most of the gold is going to get trapped. The high-grader allows you to do a quick clean-up to recover most of the gold out of your sluice, without having to clean-up the entire recovery system.

In normal production dredging, we usually clean-up our high-grader at the end of each day. Then we clean-up our entire recovery system about once every week.

High-graders are particularly helpful in double-sluice recovery systems, where the lower sluice hides the gold from your view. A small high-grade section of riffles placed just upstream of the classification screen will allow you to get a fast look at how you are doing after each sample hole.

The place you most commonly find high-graders on triple-sluices is at the head of each side box.

## A FEW OTHER POINTS ON RECOVERY

As material is being sucked up through a length of suction hose, the gold will quickly move through the material to the bottom of the suction hose, where the water and material will be moving slower because of friction. So the majority of gold will be moving slower, and will be riding along at the bottom of the suction hose as it enters the power-jet.

**DUEL-EDUCTORS AND GOLD:** On larger-sized gold dredges, a dual-eductor power-jet provides additional efficiency because there is a more orderly mixing (*venturi*) and less friction-loss inside the main jet tube (See figure 5-9).

Dual-eductor power-jet systems are efficient for another reason, too. When the jet eductors are placed horizontally, as pressurized water enters the main jet

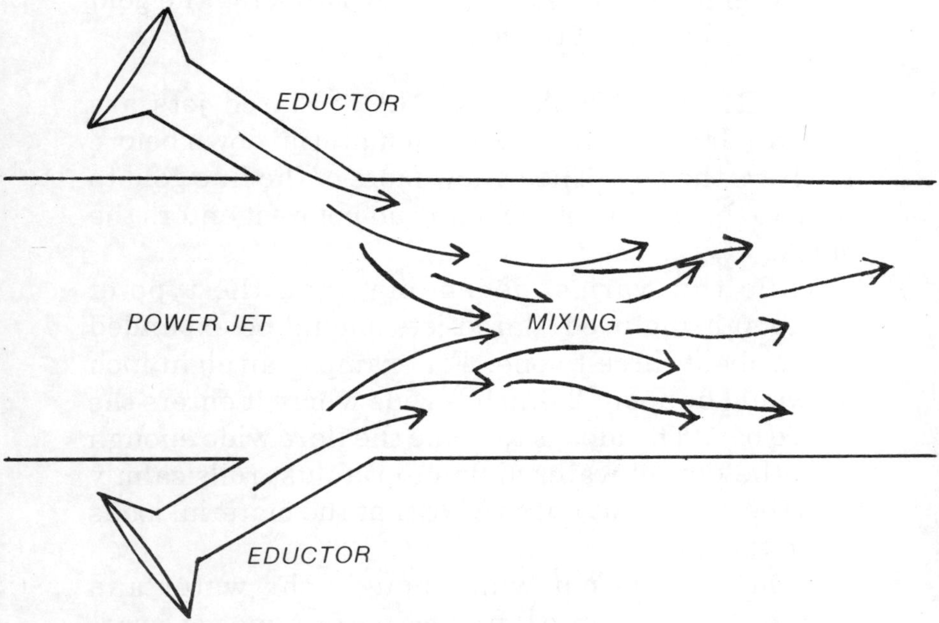


*Figure 5-9: Duel-eductors on a power-jet will provide better mixing and less friction-loss.*

tube, it will not greatly disturb the gold which is riding along at the bottom of the hose and jet. Therefore, material is allowed to enter the sluice box with the gold still riding right along at the bottom where you want it, where 20-25 feet of travel through the suction hose has placed it for you. Single eductors, triple eductors, five eductors, and couple-jet systems blow the gold up off the bottom of the jet and into the material again just before it enters the sluice box.

The proper angle of an eductor as it enters the main jet tube is about eleven degrees. This will provide optimum mixing without greatly disturbing the gold along the bottom of the jet

**FLARED JETS AND GOLD:** Flared jets are supposed to slow the water and material down before it enters the box. They allow more of the fine gold to drop to the bottom of the material before it enters the sluice.

While this varies, depending upon the type of pump(s) being used, flared jets should be expanded out to about three-to-one. For example, an eight inch jet would be about 24-inches wide where it enters the sluice box. The idea is to make the flare wide enough that the flow of water and material just rolls calmly into the sluice, but not so wide that the material loads up in the flare

If the flare is not wide enough, the water and material will flow into the sluice too fast and get swept across the classifier too quickly to do much good.

So here's the basic idea: Passage through the suction hose will place most of the gold in the slower-moving water at the bottom. Then a flared jet defuses the water flow to the point that it just calmly washes into the head of the sluice box, with nearly all of the gold moving more slowly along the bottom. The material then is swept across the first classification screen, where most of the gold can drop through and be directed to your high-grade section of riffles. A fast and easy clean-up at the end of each sample or production-run will provide you with nearly all the gold that was sucked off the bottom of the river.

**ALASKAN DAMPERS:** Long and heavy Alaskan dampers seem to work better than short ones to slow down the water flow through a sluice box. They seem to be more effective than short ones at improving fine gold recovery.

**MINER'S MOSS:** Keep your eyes open the plastic carpet underlay material that is being called *"miner's moss."* This stuff works great underneath your riffles. Most of the dredge shops are now selling it (See figure 5-10). A lot of store-bought dredges now come with it.

Miner's moss is available both with and without a rubber backing to it. It works great either way. A length of carpet or burlap should be placed underneath of the un-backed kind if it is used in a sluice box where the riffles do not bind tightly enough to the bottom of the sluice to guarantee a good seal.

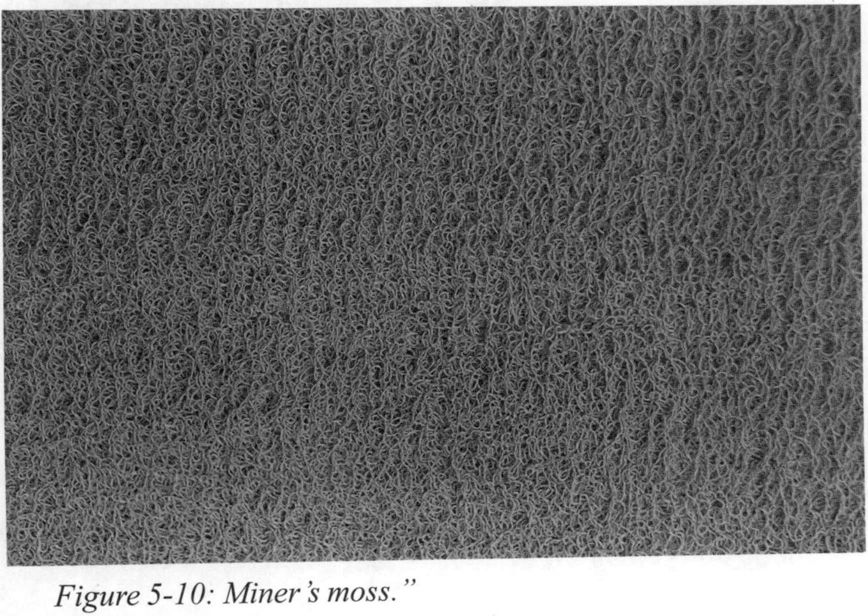

*Figure 5-10: Miner's moss."*

**SUN SCREEN RIFFLES:** Also keep your eyes open for the sun screen-type of expanded-metal. This is a taller kind of turned-up expanded-metal, which acts as a great recovery system in the side boxes of a triple-sluice, in the under-sluice of a double recovery system or underneath a classification screen in a single sluice box. Sun screen is lighter than regular expanded-metal. It comes either in aluminum or in steel. Sometimes it is hard to find. You have to call around to the various steel and metal dealers.

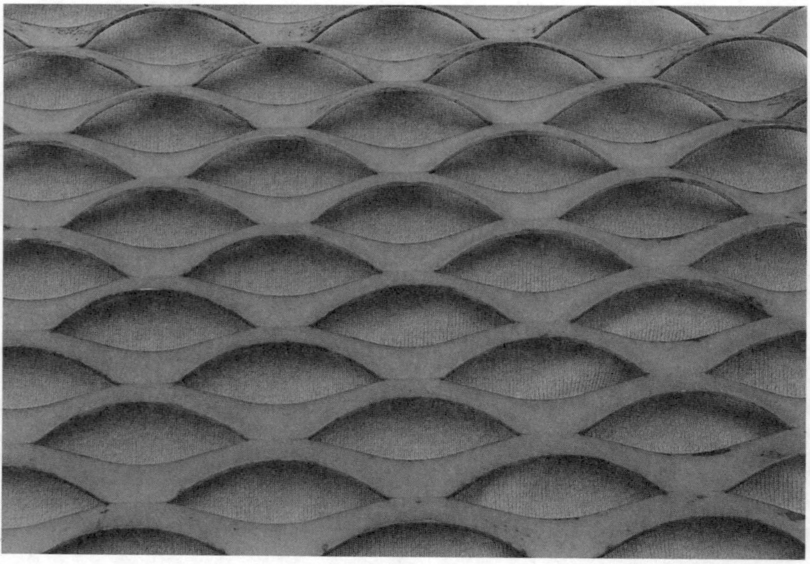

*Figure 5-11: Sun Screen-type expanded-metal makes a very good riffle system.*

# GLOSSARY

**ACCEPTABLE REQUIREMENT**: The minimum amount of gold that you must recover on an average daily basis to make a pay-streak worth dredging in a production program.

**ACCESSIBILITY**: 1) Having the legal right to bring your equipment into a location and mine it. 2) The degree of difficulty involved in getting your equipment, supplies and crew into a location.

**ALASKAN DAMPER**: A device usually made of rubber that attaches to a sluice box and rides along the surface of the water moving through the box. The *"damper"* does several things to improve the recovery of the sluice box, including smoothing out the flow and breaking the surface tension of the water as it passes through the box.

**AMALGAMATION**: The process of using mercury to recover the smaller particles of gold out of heavy concentrated material from a mining recovery system.

**ANCIENT MATERIALS**: Streambed materials that are either located in a virgin streambed or materials that have been carried from a virgin streambed during a major storm and redeposited elsewhere.

**ANCIENT STREAMBED**: For our purposes, this would include a natural streambed that has been in place since before the early miners arrived on the scene. Those gravels or an area never mined before.

**ASSAY**: An analysis of an ore sample, streambed sample or a gold sample to determine the proportions of gold, silver, platinum, and/or other metals to the amount of waste material present in the sample.

**BACK-PRESSURE**: Hydraulic forces which cause an area of water, within a waterway, to move backwards against the main flow. A back eddy is an example of a *back-pressure* area.

**BACKTRACKING**: To follow a gold deposit back through the shallow material of a pay-streak in an effort to locate the downstream boundary of the deposit.

**BAILING OUT**: When a diver drops the weights that hold him under water so he can quickly get to the surface, usually during an emergency.

**BEDROCK PAY-STREAK:** A pay-streak that is lying immediately on top of the bedrock.

**BENCHES**: Also known as *"terrace placers."* Sections of old streambed that have been left high and dry as the present waterway cut its way down to the present level. Many benches still contain large-paying quantities of gold.

**BLEEDING OFF**: The natural erosion of bench gravels into the present waterway.

**BOILING:** If you put too much water flow over top of any riffle, the suspended medium becomes so violent that even the heaviest particles can be washed away.

**BOIL-OUT:** The effect of a great force of water being moved over the top of a back-pressure area, when the force of the flow being pulled into the vacuum becomes so turbulent that even the heaviest materials, including gold, can be *"boiled out."*

**BONANZA**: A very rich gold deposit, either lode or placer, is often referred to as a *"bonanza."*

**BOULDER**: Rocks that are bigger than a large cobble. For our purposes, a *boulder* is a rock that is large enough that you must use your body weight to move it.

**BOULDER-BOUND:** A dredge hole with so many boulders at the bottom that it is almost impossible to get a boulder harness around any one of them because they are so tightly-packed together.

**BOWLINE KNOT**: A type of sailor's knot characterized by a non-slipping loop. There are important uses for this kind of knot in a dredging operation.

**CLASSIFICATION**: The process of screening out the larger-sized materials from a body of ore, streambed gravels or a set of concentrates. *Classification* is usually accomplished through the use of one or more sizes of mesh screening.

**CLASSIFIED MATERIALS**: Those materials that have been screened through a certain size of classification screen.

**CLASSIFIER**: A screening device is often referred to as a *"classifier."* There is a classifier commonly found at the head of most sluice boxes on gold dredges today.

**CLASSIFY**: See "CLASSIFICATION."

**CLEAN-UP**: The process of separating the gold from a set of concentrates, then cleaning the gold so it recovers its natural look.

**CLEAN-UP GEAR**: Any equipment that is used in the clean-up process, such as mesh boxes, gold pans, etc.

**COBBLES**: Rocks that are too large to pass through the suction-hose intake nozzle, but are smaller than a boulder. *Cobbles* have to be removed from a dredge hole by hand.

**COMMON GOLD PATH**: Because the specific gravity of gold is six times heavier than most of the other streambed materials being swept downriver during a major storm, it tends to follow and be deposited along a *common path.* For the most part, this path tends to hug the inside bends while tracking the shortest route between the bends. The first goal in a sampling program is to locate the *common gold path* within the waterway.

**CONCENTRATES**: The accumulation of heavy materials which get trapped in a gold recovery system.

**CONTACT ZONE**: That area within a streambed either where two layers come together, or where a layer rests upon the bedrock. Almost all gold you will recover in dredging will be found within a contact zone.

**CREVICE**: A split, crack, or open fissure in the surface of bedrock. *Crevices* that are located along the common gold path are likely to trap gold as streambed material is passed over them during a major flood storm.

**CLOUD-OUT**: A condition that results when the rock-person moves the wrong cobble in a dredge hole, or when material caves in from the side or working face of the hole, causing silt to wash free and obscure visibility in the working area.

**DIRECTIONAL-CHANGE BLOCK**: (Winching) A heavy-duty pulley used to winch boulders. "*Snatch blocks*" are anchored in specific locations so that boulders can be pulled in the desired direction.

**DOUBLE-BLOCKING**: (Winching boulders) A winching technique performed by attaching a snatch block to the boulder sling, then running the pull cable through that block (pulley) and back to the last directional-change anchor. This type of rigging will double the pulling force that can be exerted against the boulder by the winch and pull cable.

**DUMPING OFF**: The process of removing concentrates from a sluice box or other recovery system.

**DUST**: Very fine particles of gold, often referred to as *"gold dust."* A good gold recovery system will recover a considerable quantity of *"dust,"* often enough (or more) to pay all of the operating expenses. In the early mining days, a pinch of *gold dust* was worth about a dollar. Today, the same pinch is worth about 20-25 dollars.

**EFFECTIVE SIZE RANGE OF GOLD RECOVERY:** Any gravity recovery system will effectively recover gold down to a specific size of particle classification. The recovery system will begin losing more particles of gold when they are smaller in size than this *effective size range.*

**FAST WATER**: A condition in dredging when the water current is so swift that it will forcibly carry your body downstream if you do not take measures to prevent it.

**FAST-WATER STABILIZERS**: Extra floatation that is attached securely to each side of a dredge to increase its stability when moving it into fast water. *Fast-water stabilizers* prevent your dredge from being capsized or swamped by the swifter-moving water.

**FILLER MATERIAL**: Material that works its way down between the major rocks of a streambed as it is forming. *Filler material* can harden over a period of time and cement the rocks together to some degree.

**FINDER'S FEE**: A percentage of gold-recovery that is given to someone for their help in locating a paying deposit.

**FINE GOLD**: Particles of gold that are small enough to pass through a 40-mesh screen.

**FINENESS**: A measuring system used to indicate the purity of a gold sample or specimen. A specimen having a *fineness* of .900 would be 90% gold. A *fineness* of .650 would be equivalent to 65% gold.

**FIXED RECOVERY SYSTEMS:** Gold recovery systems that use a flow of water over riffles are called *"fixed recovery systems."*

**FLAT ROCKS**: It is not uncommon to find an increase in the percentage of *flat rocks* in a low/back-pressure area.

**FLOOD LAYERS**: Separate layers of material in a streambed that were laid down during storms occurring at different times. Each separate layer usually contains its own gold in varying amounts. The gold almost always concentrates at the bottom of a *"flood layer."*

**FLOUR GOLD**: Same as *gold dust*. (See DUST.)

**FLOURED MERCURY**: Mercury that has been broken down into thousands of tiny droplets.

**FORWARD-TRACKING**: To dredge a narrow path forward (in an upstream direction) towards the upper-end of a pay-streak, following the line that seems to pay the best, in an effort to find an extension of the pay-streak if it is there.

**FREE GOLD**: Those particles of gold in an ore or in a set of concentrates that are not chemically bonded to the other elements.

**FULL FACE MASK**: A full-sized diving mask which covers the entire face. *Full face masks* can be used in cold-water diving to prevent your face from being exposed directly to the cold.

**GENERAL LOCATION**: The larger, general area where a miner chooses to conduct his mining operation.

**GRAIN**: 1) A term used to refer to a small particle of gold. 2) A unit of weight in the troy system of weight measurement: 24 *grains* equal 1 pennyweight. One *grain* equals 64.8 milligrams.

**GRUBSTAKE**: The credit extended by a storekeeper, or the money put up by an investor, to outfit a prospector and keep him supplied until he makes a strike. In exchange, the person who funded the program will usually receive an agreed-upon percentage of the gold recovery.

**GUT**: The river's channel is often referred to by dredgers as the *"gut."*

**HARD-PACK**: Natural streambed, formed by major flood storms, is almost always tightly packed together in a structure that must be taken apart like a puzzle. Almost all pay-streaks will be found to have at least one layer of hard-pack resting on top.

**HARD PAN**: Also called a *"contact zone."* Where the bottom of a flood layer and the top of the layer just below come together. The layer just below the flood layer is usually hard-packed enough that the later flood was not able to wash it away, leaving it there as a false bedrock. *"Hard pan"* is the contact zone between streambed layers. It is where most of the gold will have concentrated if it moved with the upper streambed layer.

**HEADER BOX**: The forward-most section of some sluice box designs on dredges. A *header box* slows down the streambed materials and spreads them out evenly over a classifier as they enter the sluice.

**HEMATITE**: An iron ore ($Fe_2O_3$) which is reddish-brown in color. *Hematite* is one of the principal nonmagnetic minerals found in black-sand concentrates.

**HIGH BENCH**: See BENCHES.

**HIGH-GRADE MATERIAL**: The richest portions of streambed contained in a pay-streak.

**HIGH-GRADING**: 1) The act of stealing the higher-grade gold specimens (or a portion of the gold recovery) from a mining operation, usually by one or more of the workers. 2) Also, the practice of mining only the richest portion of a pay-streak.

**HOOKAH AIR SYSTEM**: A breathing-air delivery system used on most gold dredges, which pumps the air from a compressor on the surface through an extended airline to the divers down below.

**HOT-WATER SUIT**: A protective suit used to keep cold-water divers warm by means of pumping water, which has been heated at the surface, down through a hose and into the suit. With a continuous supply of warm water, even an old wetsuit can be used to keep a diver comfortable when dredging in cold water.

**HYDRAULIC CLASSIFICATION SYSTEM**: A means of classifying streambed materials by a recovery system using water (hydraulics) to wash the materials across classification screens. Double and triple-sluices are examples of *hydraulic classification systems.*

**HYDRAULIC MINING**: *(Hydraulicing)* A method of gold mining by which a large volume of water, under pressure, is directed by hoses to dislodge gold-bearing streambed materials located on the benches and hillsides above the level of the existing waterway. The pressurized water stream erodes the gravel down through sluice boxes, where the gold can be recovered.

**INTERMEDIATE**: A term used to identify medium-sized dredges. *Intermediates* usually include dredges of the 4-inch and 5-inch suction hose size. *Intermediates* are light enough that they can be moved around without too much trouble. Yet, they can also process a large enough volume of material to pay well when used to develop a high-grade pay-streak.

**JET**: *(Power-jet)* A tubular component on a suction gold dredge through which water is pumped into the upper portion of the suction hose, thus creating the vacuum that provides the suction-power at the nozzle. A *jet* is designed so that streambed material can be pumped directly into the recovery system without passing through the water pump.

**LAYERS**: Over a period of time, different major flood storms deposit new streambed material on top of older material. These different deposits are called *layers*. They can usually be recognized by their different colors and degrees of compactness.

**LOCKED-IN VALUES**: Valuable minerals (gold) that are chemically attached to other elements within the black sands or an ore, requiring more-involved methods of processing so the values can be released and recovered.

**LODE**: Veins that contain valuable minerals are referred to as *"lodes."* Gold, in hard-rock form *(lode)*, is commonly associated with quartz veins that protrude through the general country rock that makes up the earth's crust. *Lodes* are the original source of placer gold deposits.

**LOOSE-PACKED MATERIAL**: Material that is not tightly packed together, but is loosely situated. Often, when dredging a dredge hole in *loose-packed material*, the sides of the hole keep sliding in because there is no structure to hold the material up.

**LOOSE-PACK**: See "LOOSE-PACKED MATERIAL."

**LOW-GRADE MATERIAL**: Streambed material that contains lesser quantities of valuable minerals (gold).

**LOW/BACK-PRESSURE AREA**: Any area in a waterway where the water force slows down or reverses direction due to some sudden change in the bedrock, the presence of some obstruction, a change in the direction of flow or a widening of the channel. Gold tends to concentrate in the low/back-pressure areas that lie along the common gold path. *Low/ back-pressure areas* become pay-streaks when gold is present.

**MAGNETITE**: ($Fe304$) The heavy, magnetic black sands (iron oxide) that are found in concentrates that are collected by the recovery system.

**MAJOR STORM AND RUNOFF**: *(Major flood storm)* An incredible amount of water force is required to pick up and redeposit the streambed and gold found along the bottom of a waterway. Winter storms, in most cases, do not have the power to do this. Once in a while, however, a storm or snow runoff of huge magnitude will transport and redeposit the streambeds of an entire area. We call these *"major storms and runoffs."*

**MATERIAL**: *(Streambed material)* A term used to include all of the rocks, sand, gravel, mud, clay, silt and other sediments that make up a streambed.

**MECHANICAL RECOVERY SYSYEMS:** Recovery systems that use mechanics to help create a suspended medium are called *"mechanical recovery systems."*

**MERCURY**: *("Quick silver")* A heavy, silvery-colored liquid (at normal temperatures) metal that has a tremendous affinity for gold, silver and certain other metals. Mercury can be used to collect the fine-gold values from a set of heavy concentrates collected by a recovery system. This procedure is called *"amalgamation."*

**MESH**: A means of describing the size of a classification screen by denoting the number of openings contained in a linear inch of screening. For example, a *10-mesh* screen will have 10 openings per linear inch and 100 openings per square inch.

**NITRIC ACID**: ($HNO_3$) A clear, fuming, highly-corrosive liquid/acid that can be used to clean gold of various impurities.

**NITPICKING**: A frustratingly-slow pace of dredging caused by a dredge hole that is too small, or a nozzle-operator's production-focus being too narrow.

**NOZZLE-PERSON**: *(Nozzle-man)* The dredge operator who controls and operates the suction nozzle is normally referred to as the *"nozzle-person"* or *"N/P. "*

**OBSTRUCTION**: A rock or group of rocks that cause a plug-up in a suction hose or power-jet.

**OVERBURDEN**: The lower-grade streambed material that lies on top of a pay-streak. In dredging, the *overburden* must be removed first, before recovery of the pay-streak is possible.

**PACKING-UP:** is when you overwhelm a riffle with too much weight of material (rocks, sand, silt), and the suspended medium is lost.

**PAY**: The amount of gold that a lode or streambed is yielding to a mining activity is often referred to as how much it is *"paying."*

**PAY-LAYER**: See "CONTACT ZONE."

**PAY-STREAK**: A *pay-streak* is a concentration of gold that has formed in a section of river or stream, usually due to the slackening of the water force in that area during a major flood storm.

**PENNYWEIGHT**: A unit of the troy system of weight measurement. One *pennyweight* equals 24 grains, or 1/20th of a troy ounce.

**PLACER DEPOSIT**: Free gold that has eroded away from its original lode and been swept into a stream of running water will tend to accumulate in certain predictable locations. These accumulations are called *"placer deposits."*

**PLATINUM**: A family of 6 rare and valuable metals that are usually silvery-white or steel-gray in color. Since *platinum* is sometimes recovered along with gold (it is heavier than gold, with a specific gravity of 21.45, compared to gold's 19.3), the placer miner should know what it looks like so he does not discard it along with the waste materials from his or her recovery system.

**PLAYED-OUT**: All gold deposits, whether placer or lode, can be mined to exhaustion. When all of the known significant quantities of gold or other valuable minerals have been mined out of a deposit, it is said to have *"played-out."*

**PLUGGER POLE**: (also called a *"jam rod"*) A long, thin steel rod that is applied from the surface to tap out the plug-ups in the power-jet of a dredge.

**PLUG-UP**: A *plug-up* is an obstruction caused by a rock, or combination of rocks, that lodges in the suction hose or power-jet of a dredge and prevents the passage of material.

**POKE**: A miner's container of gold. In the early days of mining, a leather bag was often used to hold a miner's collection *(poke)* of gold. Today, plastic and glass jars are most commonly used for this purpose.

**POWER-JET**: See "JET."

**PRIMER**: A device on a suction dredge that connects the water pump to the water. A *primer* is sometimes designed to be manually filled with water so the pump can be easily primed from the surface.

**PROSPECTING**: The process of locating a deposit of gold or some other valuable mineral. *Prospecting* usually involves testing separate samples of material taken from different locations, while tracing positive signs to locate and follow the general gold path until an acceptable deposit is identified, at which time mine-development activities (production) can begin.

**PUMPING MATERIAL**: The action of feeding streambed material into the suction nozzle. The material flows up through the suction hose and is directed across the sluice box, where the gold can later be recovered from the concentrated material.

**PUNCHING HOLES**: See SAMPLING.

**RECOVERY SYSTEM**: That part of a mining operation, or a piece of specialized mining equipment, that is designed to recover the gold from the material that is directed through the system. The *recovery system* on most dredges consists of some hydraulic classification and one or more sluice boxes.

**RICH**: Streambed material that contains high-grade quantities of gold is often said to be *"rich."*

**ROCK-PERSON**: *(Rock-man)* When two (or more) dredgers are working together under water, one will control/operate the nozzle (nozzle-person), while the other one (the *rock-person* or *R/P*) moves rocks out of the way. The *rock-person* contributes to the dredging operation by removing obstructions and distractions so the nozzle operator can direct more volume of streambed material up the nozzle.

**ROLLING HITCH**: (Winching) A *rolling hitch* is set up by slinging a boulder backwards, and then running the chain or pull cable over the top of the boulder. This will usually free the boulder up by rolling it.

**ROYALTY**: A payment or commission. It is defined as a fixed percentage of the gross or net recovery. A claim owner usually receives some percentage of what an operation recovers from his claim. A grubstaker or the owner of the dredging equipment may also receive a *royalty*.

**SAMPLE HOLE**: Dredge prospecting usually involves dredging a series of *sample holes*. The first ones are intended to find the common gold path. Additional *sample holes* are dredged along the gold path in search of a pay-streak.

**SAMPLING**: Exploratory testing of materials conducted to locate high-grade gravels prior to starting a production-mining operation. This testing is usually accomplished by dredging *sample* holes. The material from each sample hole is processed through the dredge to determine the gold content. The dredger continues to excavate *sample* holes until a sufficient pay-streak is located. Then production-mining activities are commenced to recover the gold from the deposit.

**SCUBA**: *Self-Contained Underwater Breathing Apparatus* (e.g., SCUBA tank, etc.) An underwater air-breathing system (supported by tanks of compressed air) that was used during the early days of suction dredging, but which has since yielded to HOOKAH air-breathing systems.

**SILVER**: A shiny, gray-white precious metal (specific gravity 10.5) that is usually found in close association with gold. *Silver* is seldom found as a "free-standing" metal in placer deposits. It is usually alloyed with the gold recovered from the deposits.

**SINGLE DEPOSIT**: A *single deposit* of gold can form anywhere along the path that it travels in the river, usually where an obstruction or an irregularity or crevice in bedrock is located. The small amounts of gold in *single deposits* are part of the trail of gold along the common path that it follows in the waterway.

**SLOPE**: The term commonly used to refer to the downward gradient of a sluice box or waterway.

**SNATCH BLOCKS**: (Winching) A heavy-duty pulley that can easily be attached to or detached from a cable.

**SPECIFIC GRAVITY**: The weight-ratio of a volume or mass of any substance, compared to the weight of an equal volume of pure water. The *specific gravity* of gold is 19.3. This means that any volume of pure gold will weigh 19.3 times as much as an equal volume of pure water.

**STREAK**: Same as PAY-STREAK.

**SWEPT BEDROCK**: Bedrock that is consistently exposed to a force of water flow that is too great to allow very much streambed to form on top of it. It is clean bedrock. There is little or no hard-packed material sitting on top of it.

**TAIL-END**: The lower-end of a pay-streak where the streambed material no longer produces an acceptable recovery of gold values.

**TAILINGS**: The streambed material which has already passed through a dredge's recovery system and been discarded to re-fill the dredge hole.

**TAIL RACE**: (historical) Usually a channel cut into bedrock where tailings were directed after being discarded from the recovery system of a surface-mining operation.

**TROY WEIGHT**: A system of weight measurement that is commonly used by miners and gold dealers. The most common units of weight within this system are:

24 grains . . . . . . . . . . . . . . .= 1 pennyweight
20 pennyweight . . . . . . . . . = 1 troy ounce
12 troy ounces . . . . . . . . . . = 1 troy pound

**ULTRA-FINE GOLD**: This is a generalized term referring to gold that is very fine in particle size. See DUST.

**UP-CROPPING**: An extension of bedrock (usually tilted on end) that protrudes up and out of the foundation of a waterway. *Up-croppings* can be the cause of single deposits or pay-streaks when they lie along the common gold path.

**VALUES**: Those minerals or metals that are contained within streambed materials or concentrates that are of *value* to the gold dredger.

**VIRGIN GROUND**: An area that has never been mined before.

**VIRGIN STREAMBED**: A naturally-formed streambed in a location that has never been mined before by anyone.

**WATER DAM**: A barrier in a recovery system, below the classification screen, that slows the water down so that fine-gold recovery may be improved.

